99 Tricks and Traps

for

Oracle Primavera P6

PPM Professional

The Casual User's "Survival Guide"
Updated for Version 21

By

Paul E Harris

of

Eastwood Harris Pty Ltd

99 TRICKS & TRAPS FOR ORACLE PRIMAVERA P6 PPM PROFESSIONAL

Oracle and Primavera are registered trademarks of Oracle and/or its affiliates.

Windows, Microsoft® Project 365, Microsoft® Project Standard 2021, Microsoft® Project Professional 2021, Microsoft® Office Project Standard 2019, Microsoft® Office Project Professional 2019, Microsoft® Office Project Standard 2016, Microsoft®Office Project Professional 2016, Microsoft® Office Project Standard 2013, Microsoft®Office Project Professional 2013 Windows and Excel are registered trademarks of Microsoft Corporation.

Elecosoft Powerproject is a registered trademark of Elecosoft®.

Adobe® and Acrobat® are registered trademarks of Adobe Systems Incorporated.

All other company or product names may be trademarks of their respective owners.

Screen captures reprinted with authorization from Oracle Corporation.

This publication was created by Eastwood Harris Pty Ltd and is not a product of Oracle Corporation.

DISCLAIMER

The information contained in this publication is to the best of the author's knowledge true and correct. The author has made every effort to ensure accuracy of this publication but may not be held responsible for any loss or damage arising from any information in this publication. Furthermore, Oracle Corporation reserves the right in their documentation to make changes to any products to improve reliability, function, or design. Thus, the application of Service Packs or the use of upgraded software may result in the software operating differently from the descriptions in this publication.

AUTHOR AND PUBLISHER

Paul E Harris
Eastwood Harris Pty Ltd
PO Box 4032
Doncaster Heights, 3109
Victoria, Australia

e-mail: harrispe@eh.com.au
Web: http://www.eh.com.au
Tel: +61 (0) 4 1118 7701

Please send any comments on this publication to the author.

ISBN 978-1-925185-90-4 - paperback

ISBN 978-1-925185-91-1 - eBook

3 June 2022

99 TRICKS & TRAPS FOR ORACLE PRIMAVERA P6 PPM PROFESSIONAL

CURRENT BOOKS PUBLISHED BY EASTWOOD HARRIS

Planning and Control Using Oracle Primavera P6 Versions 8 to 21 PPM Professional

Oracle Primavera P6 Version 16 EPPM Web Administrators Guide

Planning and Control Using Oracle Primavera P6 Versions 16 EPPM Web

Create and Update an Unresourced Project Using Elecosoft (Asta) Powerproject Version 16

Planning and Control Using Microsoft Project 2013, 2016 and 2019

Planning and Control Using Microsoft Project 365 - Including Microsoft Project 2010, 2013, 2016 and 2019

99 Tricks and Traps for Microsoft® Project 2013, 2016 and 2019

Planning and Control Using Microsoft® Project 2013, 2016 and 2019 and PMBOK® Guide Sixth Edition

规划和控制Oracle Primavera P6 应用 版本 8.1-15.1 PPM 专业版

SERVICES OFFERED BY EASTWOOD HARRIS PTY LTD

Eastwood Harris specializes in setting up and running project controls systems with a focus on Primavera Systems and Microsoft Project software; we offer the following services:

Project Planning and Scheduling Training Courses using Oracle Primavera P6, Asta Powerproject and Microsoft Office Project:

➤ Eastwood Harris offers one-to-one training to get your new schedulers up and running quickly, without the delay of waiting for the next course and at the same time building up your own project schedule.

➤ We also run in-house training courses on any of these software packages. This is a very cost-efficient method of training your personnel.

➤ We are able to assist you in setting up a scheduling environment. This includes designing coding structures, writing procedures, training and other implementation processes.

➤ Eastwood Harris can write specialized training material that will incorporate your organization's methodology into the Eastwood Harris training manuals and develop student workshops tailored to your requirements. Project personnel will be able to use these books as reference books after the course.

Selection and Implementation of Project Management Systems

➤ Eastwood Harris will assist you by conducting an internal review of your requirements and match this requirement analysis against the functionality of packaged software.

➤ We are then able to assist you in the implementation of these systems, including writing policies and procedures and training personnel, to ensure a smooth transition to your new system.

Dispute Resolution

➤ Eastwood Harris is able to analyze your subcontractor's schedules in the event of claims and provide you with a clear picture of the schedule in relation to the claim.

Schedule Conversion

➤ Eastwood Harris is able to convert your schedules from one software package to another. The conversion of schedules is often time consuming, so let us do it for you.

Please contact the author for more information on these services.

TABLE OF CONTENTS

1 INTRODUCTION

1.1 Background Knowledge required to use this book

Readers of this book should understand:

❖ How projects are managed and the processes associated with running projects,

❖ Critical Path theory including understanding Early dates, Late dates and Float calculations, and

❖ The basic functions of Oracle Primavera P6 PPM Professional.

1.2 P6 Versions

This book is applicable to Oracle P6 PPM Professional Version 7 and later.

❖ The main difference between Version 7 and 8 was the toolbars became editable in Version 8.

❖ The main difference between Version 6 and 7 was a major change to calendars. Version 7 now calculates the duration in days correctly for calendars with a different number of hours per day.

❖ There were no P6 Versions 9 to 14 because after Oracle purchased Primavera Systems, they adopted the Oracle convention of naming versions by their year of release.

❖ There are minor updates to the software with each new release but no major changes to the software.

1.3 Aim of this book

P6 has many functions and terms that are not intuitive and the aim of this book is to give P6 users a comprehensive reference book allowing them to set up a database and use P6 with confidence.

1.4 Information Sources

This book has a large amount of information taken from my other books, some of which I have summarized. I have also included several techniques from other sources thus providing some more advanced information than my other P6 books.

1.5   Important Points

Primavera P6 has functions that that do not work in an expected way and catch out users. Thus, using P6 becomes a career shortening exercise when database administrators and users do not understand these functions. Every P6 user and Administrator must understand the functions listed below and be able to identify when they should be used carefully or not be used at all:

❖ Schedulers should always display the date and time when manipulating a schedule. These are set in the **User Preferences**. If the time is not displayed then P6 may use an irrelevant time when **Actual Dates**, **Constraints**, **Suspend** dates and **Resume** dates are being set.

❖ What the **Planned Dates** are, how they are recorded by P6 and when they are displayed. It is **CRITICAL** that every P6 user and P6 database administrator understands the **Planned Dates** in both the **Current Schedule** and a **Baseline** schedule.

❖ Schedulers should understand what dates and costs are being read from a Baseline. The data that is read for all projects and all users is set in the **Admin**, **Admin Preferences**, **Earned value** tab, **Earned value calculation** drop down box. Different databases may have different Admin settings and this may result in Baselines displaying different Baseline data when a schedule is imported into a different database.

❖ When a user closes one project and opens another the original **Layout** will still be applied to a project. This causes problems when there are multiple customer projects' in one database with different customer logos and text in the customers' **Layouts**. There should be procedures and check lists written and used to prevent a schedule being presented to a customer with another customer's logo and information in headers and footers.

❖ The non-working time displayed in all Gantt Charts is adopted from the **Enterprise, Calendars, Global Default** calendar. Different projects may not display different non-working times in the Gantt Chart because all users are displayed the same **Default Calendar** non work time in the background.

❖ The **Tools, Update Progress** does work in the same way as other products with a similar named function. In P6 the use of this function may change the user assigned **Actual Start Date** and **Finish date** of in progress activities and this function should be used with extreme caution. The author recommends that users should **NEVER** use this function.

❖ There are several issues with the Oracle Primavera P6 Activity Layout bars settings supplied with Oracle databases and the **Default** bar setting that are set when the **Format, Bars, Default** button is activated. These issues must be understood. It is recommended that:

➢ The **Format, Bars, Default** button should never be used.

➢ The administrator must edit all Oracle Primavera P6 Activity Layout bar settings in line with the authors recommendations, or

➢ Download the **www.primavera.com.au_Layout** from one of the authors websites, or

➢ Set the bars formatting in all existing Activity Layouts using the **Bars** form **Copy from** 🗐 Copy From... function and copy from another layout such as the **www.primavera.com.au_Layout**.

2 GETTING THE ENVIRONMENT RIGHT – SETTING DATABASE OPTIONS

2.1 Understanding Databases

Oracle Primavera P6 may open three types of databases:

❖ **EPPM, Enterprise Portfolio Project Management.** This type of database may be opened with either the **PPM Windows Client** or the **EPPM Web Client** and the administration of the database is only undertaken in the **EPPM Web Client**. Therefore, when you open an EPPM database with the Windows Client you will find the Admin menu and some Enterprise menu functions are removed and these functions are performed in the EPPM Web Client. This mode is also referred to as the **Optional Client**.

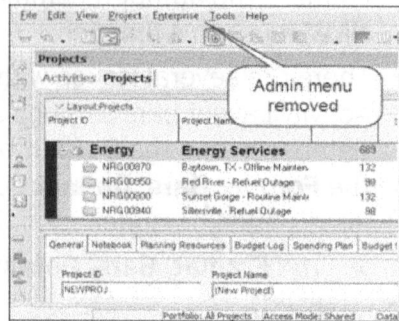

❖ **PPM, Professional Project Management.** This database may only be opened with the PPM Windows client and the administration of the database is undertaken in the Windows Client using the **Admin** menu.

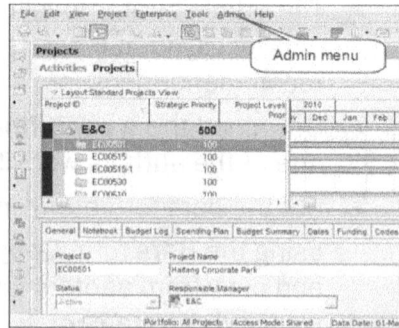

❖ There is also a **Stand Alone** load that does not have the ability to create users.

2.2   Understanding Database Options

The database options affect all users and there are many default options in both an empty database and the demonstration database that will need to be adjusted for most organizations.

This chapter will take you through the important ones that the author normally changes and these are edited using:

❖ **Admin, Admin Preferences…, General** tab in a PPM Professional database, as per the picture below, and

❖ **Administer**, **Application Settings**, **General** in an EPPM database, no pictures of the EPPM database will be displayed in this book, but these options are found under the **Admin** menu.

2.3   General

We will work through the **Admin Preferences** settings starting with the **General** tab:

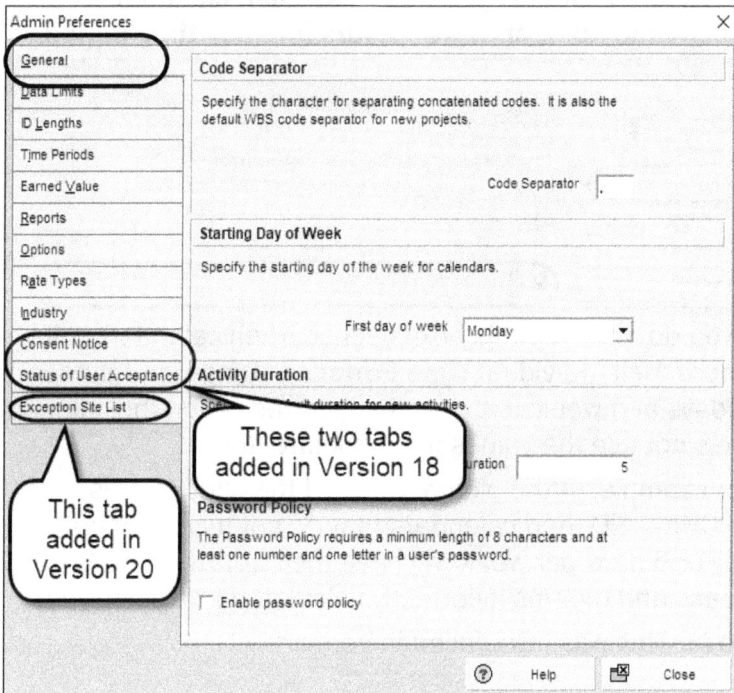

2.3.1   Start Day of Week

The **First day of week** is defaulted to Sunday. Many people prefer to see Monday because the calendar date in the weekly view is then Monday and is a workday, as per the picture below:

These dates are a Monday

2.3.2   Default Duration

This is best set at 5 days because when reduced to 1 day a new activity bars becomes hard to see, because it becomes a very short bar.

2.4   Time Periods, Hours per Time Period

ALWAYS check the option **Use assigned calendar to specify the number of workhours for each time period**.

If you do not keep this box checked, then calendars will ignore their individual **Time Periods** settings, and durations in days and weeks etc. may be incorrect when the calendar does not use the values in the picture above.

For example, in the picture above if the check box is UNCHECKED then calendars that do not have 8 hours per day or 5 days per week will have their durations in days, weeks and months incorrectly calculated and displayed.

NOTE: This was introduced in Version 7.

2.5 Earned Value, Earned Value Calculation

Like many descriptions in P6 this does not mean exactly what is the title suggests. These options decide which Baseline schedule values are read to:

❖ Calculate the **Earned Value** fields in the current schedule, and

❖ Which baseline dates are displayed as **Baseline** bars.

A baseline has two sets of data for dates, costs and units that may be read and displayed as baseline data. An administrator must decide what data all current schedules will read and ensure users understand the setting:

❖ **Dates** – will a baseline display the:

➢ **Planned Start** and **Planned Finish**, or

➢ **Start** and **Finish**

❖ **Costs and Units** - will a baseline display the:

➢ **Budget** (or **Planned** in some industry versions and the EPPM Web tool)

➢ **At Completion**

The **Planned Start, Planned Finish** and **Budget** or **Planned** may hold irrelevant data. See the Planned Dates section 15.4.

The **At Completion values with current dates** is the author's preferred option when resources are assigned.

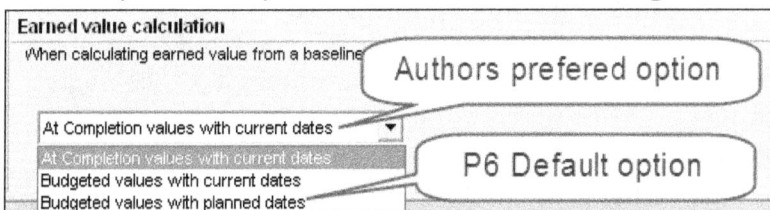

NOTE: These options apply to every user. Also, when you import a Baseline schedule from another database you should check what options the other database has selected. Not understanding how these options work may be a career shortening exercise!

2.6 Options

The options in the **Options** tab from increase over time as per the picture below:

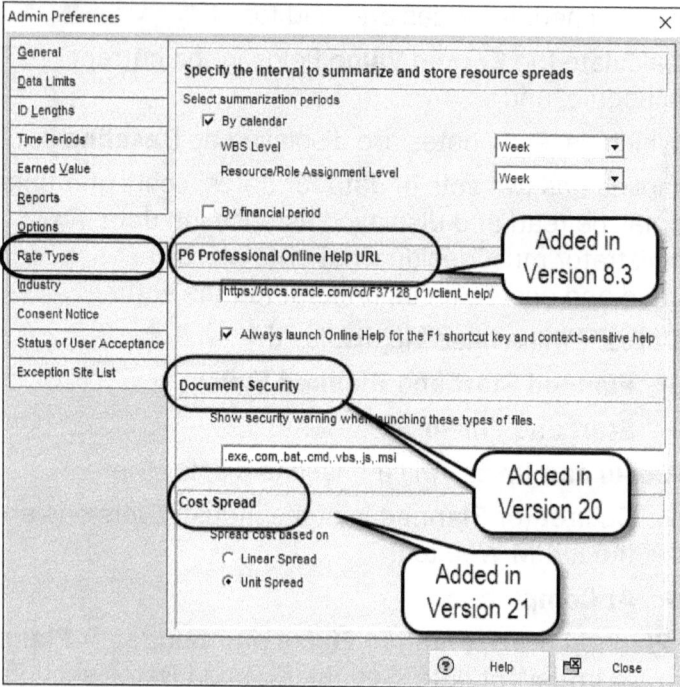

2.6.1 Specify the intervals to summarize and store resource spreads

The **Projects** and **Tracking** windows do not read the latest current schedule data. They read **Summarized Data** which is updated when a database is summarized, or the project is open.

❖ The level at which this data is stored in the database is set in this tab and weekly is usually the most suitable.

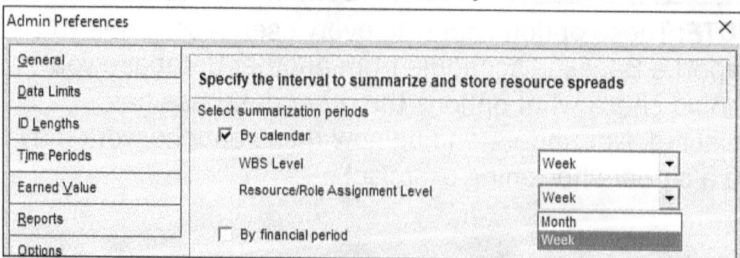

❖ A database is summarized using **Tools, Summarize.** Once summarized the data read in the **Projects** and **Tracking** windows will be correct at the time of summarization.

➢ A database running from a server should be summarized every night by setting up a **Job Service** using the **Tools, Job Service** command.

➢ A database running on a standalone database should be summarized using **Tools, Summarize** before any data is read in the **Projects** and **Tracking** windows.

❖ The WBS level that the data is summarized to (and when the data was last summarized) is set by the user per project in the **Project** window, **Settings** tab, **Summarized Data**. This sets the level of detail available in the **Tracking** window.

Settings	
Summarized Data	
Last Summarized On	
14-Jan-19 16	
Summarize to WBS Level	
2	⬍

2.6.2   Options Removed from the Options tab

The **Project Architect** and **Web Access Server** URL check boxes were removed from the Professional Client.

Enable Link to Contract Management Module (originally called Expedition) enables linking to this module when installed, removed in Version 19.

2.6.3   Professional Online Help URL

Thus sets the URL for accessing help and should not normally be changed, unless your company has produced its own help files.

2.6.4   Document Security

In P6 Version 20 You may prevent people from downloading harmful files in the **Admin Preferences, Options, Document Security** by listing the file types users may not download.

2.6.5 Cost Spread

❖ P6 Version 21 introduced **Cost Spread** which gives a more accurate cash flow over the change in rate of both resources and roles and is covered in the **Creating Roles and Resources** section 19.3.11.

2.7 Admin, Rate Types, Resource and Role Rate Types

Primavera has five **Resource and Role Rate Types** which may be renamed in this tab. This allows each role or resource to be assigned up to 5 rates. For example, there could be an internal and external charge out rate for resources:

Most construction companies create a set of resources for each project with a rate for that project and only use the first Resource Rate. In this situation it is recommended that all the other four **Resource and Role Rate Types** are renamed **Do Not Use**.

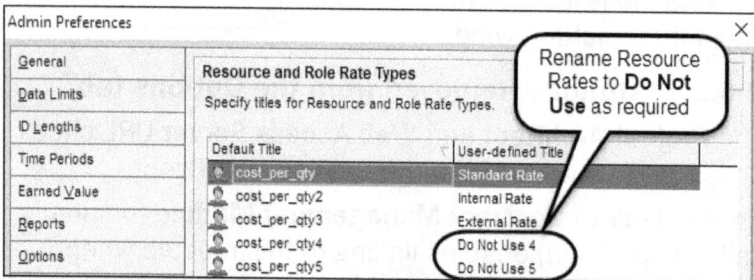

When a project is created the **Default** rate is selected in the **Project** window, **Resources** tab, **Assignment Defaults** section:

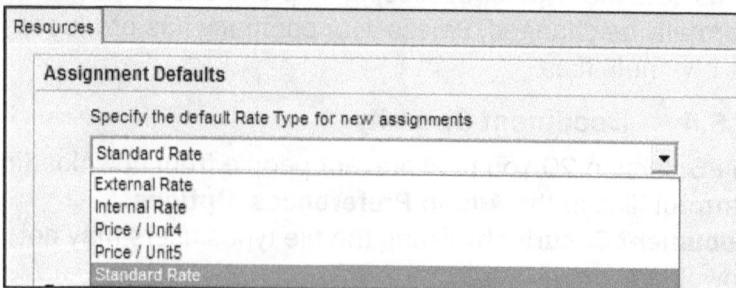

The rate used by resources assigned to an activity may be changed at any time in **Activities** window, **Resouurces** tab, **Rate Type** drop down box:

2.8   Admin, Preferences, Industry

The **Industry type** determines the terminology used in some fields and in earlier versions was set when the software was loaded. This now may be set in the P6 Professional by selecting **Admin**, **Admin Preferences...**, **Industry** tab:

When an EPPM database is being used then the Industry is set with the Web Client in the **Administration**, **General**, **Industry Selection** drop down box.

The following table displays the terminology:

Industry Type	Terminology	Name of Project Comparison Tool
Engineering and Construction	Budgeted Units & Cost Original Duration	Claim Digger
Government, Aerospace, and Defense	Planned Units & Cost Planned Duration	Schedule Comparison
High-Technology, Manufacturing and Other Industry	Planned Units & Cost Planned Duration	Schedule Comparison
Utilities, Oil, and Gas	Budgeted Units & Cost Original Duration	Claim Digger

Engineering and Construction:

Government, Aerospace, and Defense:

If a different Industry Type is selected, then P6 has to be restarted to see the changes. The EPPM Web Client only uses Planned Units & Cost, Planned Duration and Schedule Comparison.

2.9   Admin, Consent Notice

Consent Notice were introduced in P6 Version 18 and alert users to any corporate policies designed to protect personally identifiable information that may be stored or transmitted when using P6.

A Consent Notice is displayed when a user first operates one of the functions enabled as a Consent Notice. The user must accept the consent notice before being allowed to progress.

2.10   Admin, Status of User Acceptance

Status of User Acceptance shows how many Consent Notices have been displayed and accepted by users.

The **User Preferences**, **Personal Information** tab allows users to see the personal information entered by the administrator when the user was created.

2.11 Exception Site List

In P6 Version 20 a list of web sites may be added to **Admin Preferences**, **Options**, **Exception Site List**.

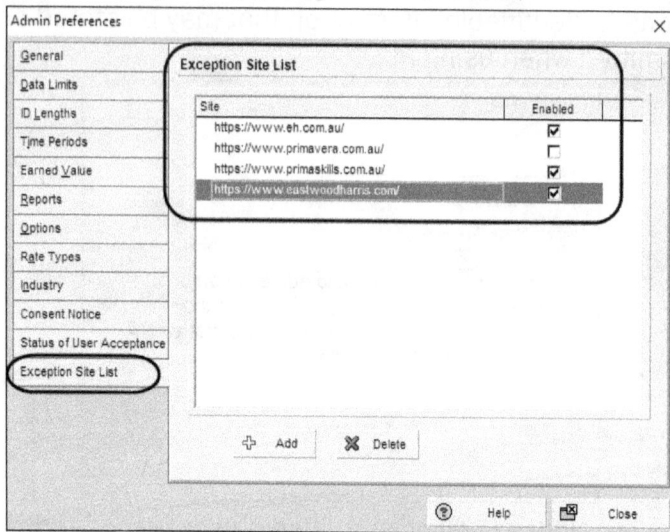

The URLs may then be added to **Notebook Topics** and UDFs and permitted URLs may be launched from P6 by clicking on the URL and selecting Launch.

Other URLs in P6 fields may be copied and pasted into a browser but not run from P6.

2.12 Admin, Categories

This menu is used to tailor the various categories which should align to your organization's project management procedures.

2.12.1 Notebook Topics

Notebook Topics are useful for recording information on a specific subject. It is unfortunate that they may not be displayed in columns in the **Projects** and **Activities** windows, thus it makes it difficult to share **Notebook Topics** with project team members who do not have P6 access.

You may wish to consider using **UDF** fields for recording information that you wish to share with people who do not have access to P6, so this data may be simply displayed in reports columns and printed out as required.

2.12.2   Units of Measurement

These are the units of measurement you use for Material Resources and should be tailored for the types of **Material Resources** your company uses.

NOTE: The default P6 load usually does not have any metric units and must be added.

2.13   Admin, Currencies

The **Admin**, **Currencies** form is used to edit currencies. When you are only using one currency it is advised that all the other currencies are deleted to avoid users selecting the incorrect currency.

The **Base** currency check box does not operate as expected and you will have to edit the **Currency ID**, **Currency name** and **Currency Symbol** in the first line to suit your countries currency and in turn making it the **Default Currency**.

The **Exchange Rate** is a simple method of calculating the cost in a different currency and it is not possible to vary this over time.

The user selects which currency to see all their project costs using the **Edit**, **User Preferences**, **Currency** tab.

NOTE: This is a dangerous function when multiple currencies are in use, as a user may hide the currency symbol and no one will know what currency the cost are displayed in! It is possible to have two **Currencies** with the same symbol and if a user selects a different currency then all costs displayed by the user will be converted to a different value. This option must be carefully monitored and if you do not need multiple currencies then it is suggested that you should delete them all, to avoid any possible problems. If you are using multiple currencies, then make sure that all currencies have a different sign so there is no confusion.

2.14 Admin, Financial Period Calendars

This is where the **Financial Periods** associated with **Storing Period Performance** are created.

Financial Periods have to be used when it is important to have data that reflects how much work was completed or costs spent in each period and not just averaged over the periods to date and using **Financial Periods** creates more accurate S-Curves.

Functionality was added in P6 Version 20 allowing different projects to have different Financial Periods and the function was renamed **Financial Period Calendars.**

Using **Financial Periods** requires a high level of discipline in the project team and are difficult to use:

❖ **Financial Periods** in Version 19 and earlier are created in **Admin**, **Financial Periods**, and all projects in a database have the same Financial Periods,

❖ **Financial Period Calendars** in Version 20 and later are created in **Admin**, **Financial Period Calendars**, and assigned to projects in the **Projects** window, **General** tab,

❖ They are set in a round number of months or weeks,

❖ The **Projects**, **Calculations** tab **Link actual to date and actual this period units and costs** must be checked for the **Financial Periods** to operate. This may be unchecked to manually fix up past errors, but must be rechecked before storing performance,

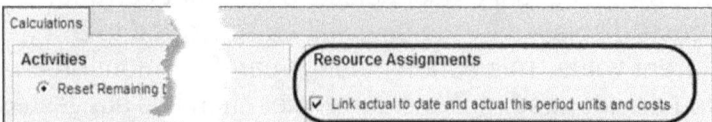

❖ Period performance is stored using **Tools**, **Store Period Performance:**

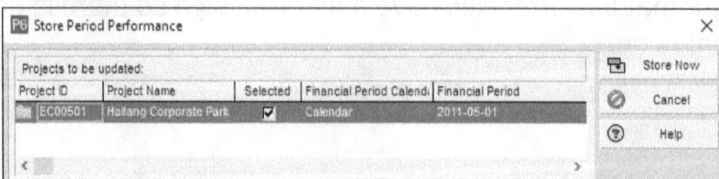

❖ Many resource windows and reports have the option of **Display Actuals using Financial Period data**.

2.15 Date and Time Display

The date and time display set in the **User Preferences** may be different for each user and are applied to any project that a user opens.

The user's date and time display are used in every window and report that the user displays.

When a company requires all project schedules to look the same, then the required format should be documented and communicated in corporate procedures.

2.16 Units Display

The **Time Units** set in the **User Preferences** may be different for each user and are applied to any project that a user opens.

This function allows different users to show different units, e.g. it allows one user to show hours and another to show days for the same project and layout.

When a company requires all project schedules to look the same, then the required format should be documented and communicated in corporate procedures.

2.17 Printouts and Reports Header and Footers

The header and footers are part of the **Activity Layout** that a user has applied to a project.

Unfortunately, when a user opens a different project the Layout from the last project will still be used to display project data in the **Activity** window. As a result, it is very simple for a user to print a report for one customer with the header and footer, including logos, of another customer.

To prevent this happening, procedures and check lists should be used when your database has multiple customer projects.

2.18 Users, Security Profiles and Organizational Breakdown Structure

This section is intended to introduce this topic. Please refer to my other books, or the Primavera Administration Manual for full details.

The full picture and processes for creating users and assigning access are:

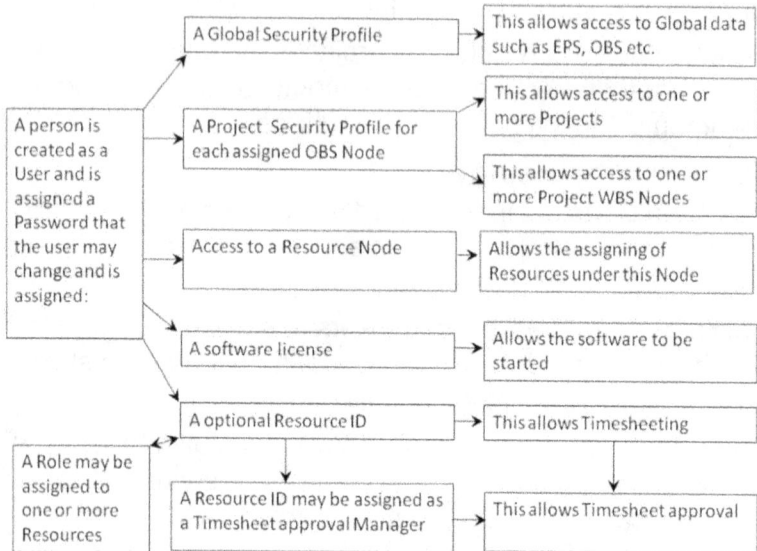

A person is created as a User and is assigned a Password that the user may change and is assigned:	A Global Security Profile	This allows access to Global data such as EPS, OBS etc.
	A Project Security Profile for each assigned OBS Node	This allows access to one or more Projects
		This allows access to one or more Project WBS Nodes
	Access to a Resource Node	Allows the assigning of Resources under this Node
	A software license	Allows the software to be started
	A optional Resource ID	This allows Timesheeting
A Role may be assigned to one or more Resources	A Resource ID may be assigned as a Timesheet approval Manager	This allows Timesheet approval

❖ The **EPS** is created, allowing projects to be created under each EPS Node. This often mirrors the company's network drive hierarchy.

❖ The **OBS** is created and acts as a security gateway for users to access projects. This may not need to represent your company's OBS and often this is set up to mirror the EPS.

❖ A user is created by selecting **Admin, Users...** and each user is assigned:

➢ A **Global Security Profile** which allows access to Global data, such as EPS, OBS, etc.

➢ A **Project Security Profile** for each assigned OBS Node, which allows access to one or more EPS Nodes, Projects, or WBS Nodes within a project.

© *Eastwood Harris*

➢ Access to all or one **Resource Nodes** is assigned to a user from the **Resource Window**. The user can only see and assign resources from this node but may see any resources and their associated costs once they are assigned to activities. P6 Version 19 introduced the ability to allow a user to be assigned up to five resources and the resources underneath the nominated five resources when defining resource access. A user may view and assign the selected resources and the child resources.

➢ Access to a software license, allowing the user to login and start the software.

➢ The user may be assigned to a resource in the **Resource Window**, thus allowing timesheets to be used.

➢ One or more **Resources** may be assigned to one or more **Roles**.

2.19   Project Codes

Project Codes are assigned to projects and enable projects to be Grouped and Sorted under an alternative structure to the EPS.

For example, when an EPS represents the physical location of offices by country, state/county and city, the Project Codes enables projects to be given tags, such as Reason for the Project, Safety, Compliance, New Product, and Increase Production. The Projects may be grouped or filtered under these headings.

Therefore, project codes are used to Group, Sort and filter Projects in a similar way that Activity Codes are used to Group and Sort Activities.

2.20   Summarizing Projects

The data displayed in the **Projects** and **Tracking Windows**, such as Durations, Dates, etc., may be incorrect unless the projects have been **Summarized** by selecting **Tools**, **Summarize**.

You will notice that the data displayed against a project in the Projects window may change when a project is opened and then at this point in time the latest data from the project is read, but when the project is closed this data is read from a **Summary Data** table in the database which is only updated when a database is **Summarized**.

The **Settings** tab in the **Project Window** specifies to what level the data is summarized and indicates when it was last summarized.

A large database takes a significant amount of time to summarize and may be summarized at night automatically using **Job Services**.

It is good practice to set up a **Job Service** to **Summarize** a database every night.

NOTE: In the picture above, selecting **Summarize to WBS Level** is set to zero so all levels of the WBS will be summarized.

© *Eastwood Harris*

2.21   Job Services

A **Job Services** may be set up in a PPM database by selecting **Tools**, **Job Services...** to open the **Job Services** form, which can perform the following functions on one or more selected projects or EPS Nodes:

Select **Administer, Global Scheduled Services** in the Web for setting up a Job Service in an EPPM database. The following functions are available:

❖ **Apply Actuals** to projects when timesheets are used.

❖ **Batch Reports**. In the **Reports Window** a **Batch** may be created by selecting **Tools, Reports, Batch Reports...** to open the **Batch Reports** form. This creates one or more reports simultaneously. A Batch may be run on a regular basis using a job service.

❖ **Export** one or more projects on a regular basis.

❖ **Schedule** one or more projects on a regular basis.

❖ **Summarize** projects. This must be set up to run nightly for all databases so the data in the projects window is correct.

3 SETTING UP USER PREFERENCES AND OTHER USER SETTINGS

3.1 Creating Users

It is recommended that the Administrator completes the following tasks when a new user is created:

❖ Ensure the user has read write access **ONLY** to the EPS Node(s) they need and do not have access to areas that they will not be working in or to functions they do not need. This will stop a lot of heartache created by the unintentional deletion of projects and data.

❖ Log in as the user to:

➢ Check their access to their projects and data,

➢ Set all the **User Preferences** in line with company project management procedures, see the next few paragraphs,

➢ Customize the menu options and disable **Menus show recently used commands first** so the user always sees all the menu options,

➢ Add the **No Bottom Layout** icon ⬓ to the **Bottom Layout** toolbar as it is always missing.

➢ Open a project and assign an appropriate Activity Layout with good bar formatting.

3.2 User Preferences

In P6 **User Preferences** are user centric. It would be better if many of the **User Preferences** were **Project Preferences**.

Also, the formatting set in the **User Preferences** apply to all views and for all projects in P6 and the user must continually change them, depending on what the user is doing. It would be better if the user could format dates and times for each Activity Layout, but this is not possible.

As a result, a user must be careful how they set their **User Preferences** and must continually change them depending on what they are doing.

3.2.1 Time Unit Formatting

It is recommended that users always display the **Sub-unit** in the **Duration** column to allow them to see if there are any non-round or whole days.

To adjust how the date and time are displayed:

❖ Select **Edit**, **User Preferences...**,

❖ Select the **Time Units** tab,

❖ The **Duration Format** section determines how the activity durations are displayed and should be set as per the picture below:

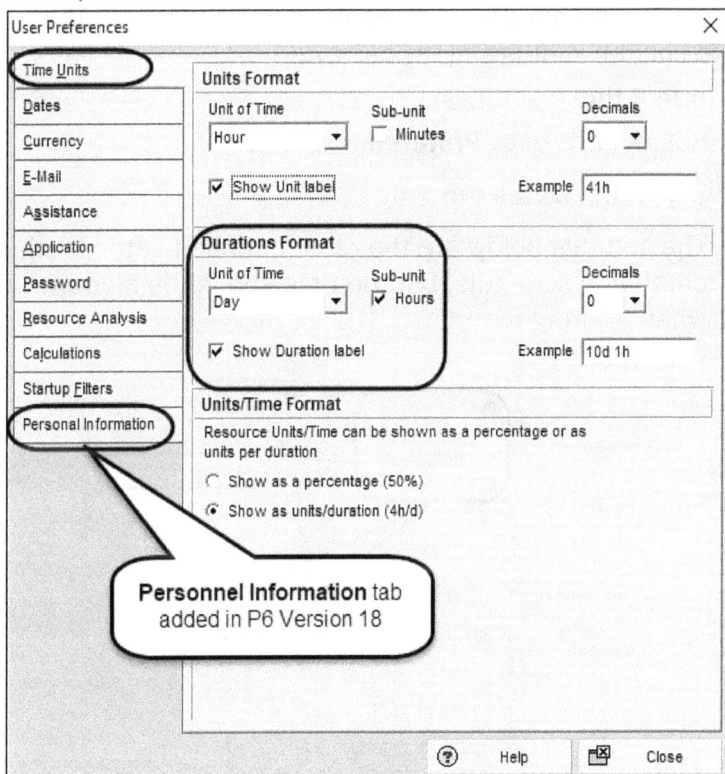

❖ The **Duration** format above is recommended by the author so any non round durations will be observed.

3.2.2   Date Formatting

Date Format:

It is recommended that users always display the **Month name** on all international projects so there is no confusion between the US mm/dd/yy format and the ROW (Rest of World) format of dd/mm/yy.

Time Units:

It is **STRONGLY** recommended that the time is **ALWAYS** displayed in 24-hour format so the user knows the time of any selected date. This is because the software will often select 00:00, (the first minute of a day) when assigning dates. The author does not display minutes to keep the date column widths slightly narrower.

To format these settings:

❖ Select **Edit**, **User Preferences...**,

❖ Select the **Dates** tab,

❖ The settings below are the recommended settings when creating a schedule, but the time should be hidden when printing reports:

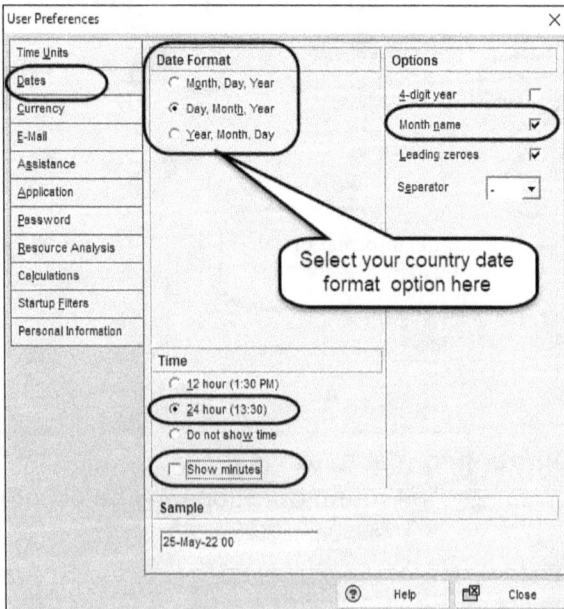

3.2.3 Users may See and Set Activity Start and Finish Times in the Date Picker box

Users may see and set Activity Start and Finish times in the **Date Picker** box when the User Preferences are set only to show the date. This is a great enhancement for P6 Version 21 as P6 usually picks the wrong time when setting Actual dates, Constraints, Suspend and Resume dates when the time is not set to be displayed from the User Preferences.

❖ P6 Version 20 and earlier the time is not available:

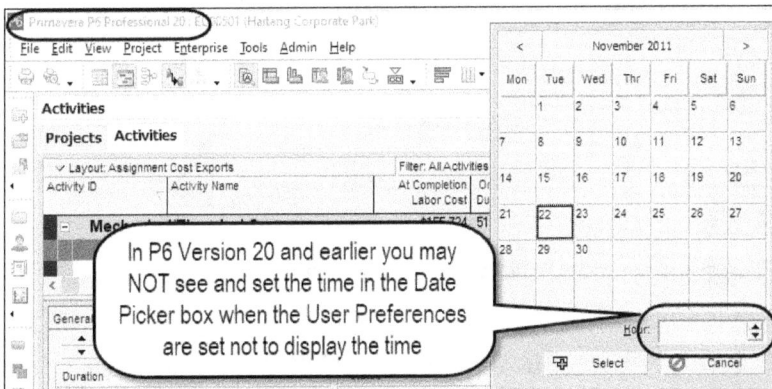

In P6 Version 20 and earlier you may NOT see and set the time in the Date Picker box when the User Preferences are set not to display the time

❖ P6 Version 21 the time may be set from the **Date Picker** box:

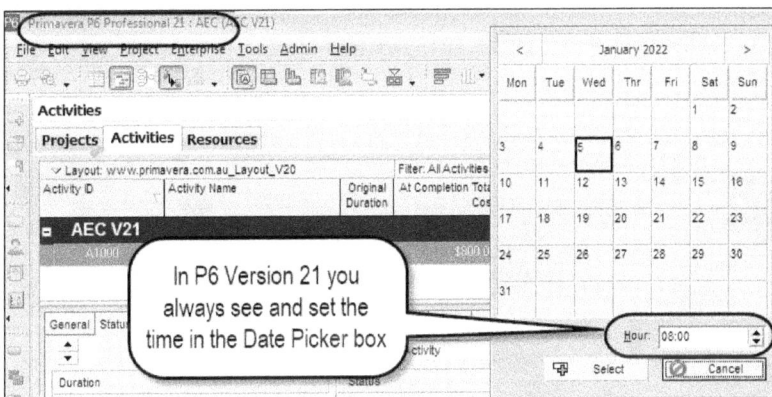

In P6 Version 21 you always see and set the time in the Date Picker box

3.2.4 Currency

The currency should be set to the appropriate currency.

It is recommended that all currencies, except your country's currency, are deleted from the database to prevent any issues with users selecting the wrong currency and have a bid submitted with the incorrect value.

User Preferences		×
Time Units	**Currency Options**	
Dates		
Currency	Select a currency for viewing monetary values	
E-Mail	🐨 Australian Dollar	...
Assistance		
Application	☑ Show currency symbol	$
Password	☑ Show decimal digits	0.00

3.2.5 Email

This section is used when a user is also a resource in the database.

3.2.6 Assistance - Stopping Wizards Running

The Wizards slows down adding activities and resources; thus, it should be disabled from the **User Preferences** form by:

❖ Selecting **Edit**, **User Preferences...**

❖ Selecting the **Assistance** tab and

❖ Uncheck both **Wizards** as shown below:

User Preferences		×
Time Units	**Wizards**	
Dates	Would you like to use wizards when adding new activities and	
Currency	resources?	
E-Mail		
Assistance		
Application	☐ Use New Resource Wizard	
Password	☐ Use New Activity Wizard	

© *Eastwood Harris*

3.2.7 Application

Application Startup Window

This specifies which Primavera window is displayed when the software is started.

❖ If you work in the same project all the time, then set this to **Activities** and also do not close the project when closing Primavera. Then, the next time you open Primavera you will be taken to your project in the **Activities Window**, thus saving time when opening P6.

❖ If you work in different projects, then select **Projects**, to be taken to the **Projects Window** when you open P6.

❖ **Show the Issue Navigator dialog at startup** and **Show the Welcome dialog at startup** should NOT be displayed as they slow down the user's access to the software.

Application Log File

This was removed from later versions of P6. When **Write trace of internal functions to log file** is checked P6 creates a log of all data entries titled ERRORS LOG.

❖ This would be used by support staff and should not be turned on unless requested by support staff.

❖ Removed in later versions of P6.

Group and Sorting

This specifies what information is displayed in the bands in forms such as the **Predecessor** and **Successors** forms that do not have **Group and Sort** formatting options. One or both options of Description or Code may be selected.

Columns

Primavera Version 5.0 introduced **Financial Periods**. This is where the user specifies which **Financial Periods** are displayed.

❖ This is a useful field when **Financial Periods** are used and there may be hundreds of columns in a schedule with many updates. This allows the user to only display the relevant ones and avoids the need to be scrolling up and down through hundreds of old irrelevant fields.

❖ On the other hand, in earlier version if you do not use **Financial Periods** then **NEVER EVER** display them, as once displayed they may not ever be hidden again and you will always have some **Financial Periods** to scroll through when setting up **Activity Filters** etc..

❖ Later Versions have an option to **Load Financial Period data** which disable the loading of **Financial Period** data and the software starts more quickly.

3.2.8   Password

The **User Password** tab is used to change the user password.

❖ An Administrator may reset a user's password at any time.

❖ A password set by a user here is not accessible by the Administrator.

3.2.9   Resource Analysis

All Projects

The **All Projects** option specifies which projects are used to calculate the **Resources Remaining Values** in **Resource Usage Profiles**.

NOTE: Projects must be **Summarized** and time intervals set to weeks for this function to operate.

Time-Distributed Data

It is possible to drag a project forward or backwards in time in the **Tracking Window** or **Portfolio Analysis**. This action creates a new set of dates titled **Forecast dates**. At this point in time you may show **Resource Usage Profile** and **Resource Usage Spreadsheet** either using the Current Schedule **Remaining Early Dates** or the revised **Forecast dates**.

NOTE: Thus, again at this point in time, different users may unwittingly display different data.

Interval for time-distributed resource calculations:

This option determines the time increment for displaying the Resource Usage Profile and Resource Usage Spreadsheet data.

❖ Selecting **Days** results in no hourly spread and all hours put at midnight:

❖ Selecting **Hours** results in a correct spread of the hours, but more memory consumed:

Display the Role Limits based on

This enables options for displaying the Role Limits in Resource Profiles.

❖ For example, a role may have been defined a limit of six resources but only have four Resources assigned to the Role.

❖ This option allows you to decide if you wish to display a limit of four based on the resources available or six based on the limit assigned to the Role.

3.2.10   Calculations

The **Calculations** tab, **Resource Assignments** section has two options:

Preserve the Units, Duration, and Units/Time for existing assignments.

❖ With this option, as Resources are added or deleted the total number of hours assigned to an Activity increases or decreases. The hours assigned for each resource are calculated independently.
NOTE: This is author's preferred option and the best option for building and construction projects.

❖ **Recalculate the Units, Duration, and Units/Time for existing assignments based on the activity Duration Type.** The total number of hours assigned to an activity will stay constant as second and subsequent resources are added or removed from an Activity, except when the Activity Type is Fixed Duration and Units/Time.
NOTE: Thus, assigning resources will reduce the work for each resource and either the activity duration will reduce or the Units per Time Period for each resource assignment will reduce. This calculation is dependent on the Duration Type.

Assignment Staffing

This allows the user to set the defaults for:

❖ Selecting the **Units per Time** when assigning a substitute resource to an existing resource assignment.

❖ Selecting the **Price per Unit** for a resource which is being assigned to a Role.

❖ These options allow the user to select the existing resource values or the new resource values when reassigning resources, or to be prompted each time a resource/role is substituted.

❖ **NOTE:** The author recommends setting it as **Ask me to select each time I assign**.

3.2.11   Startup Filters

The **Startup Filters** option in the **User Preferences** form enables the selection of **Startup Filters** when opening one of the following windows: Resources, Roles, OBS, Activity Codes and Cost Accounts.

❖ You may find when you open a window, such as the **Resources Window**, that it has no data is displayed. This is normally due to the settings in this tab and also not having any resources assigned to any activity.

❖ **NOTE:** The author recommends selecting **View all data (No Filter)** so you will not end up with a blank screen when you open windows such as the **Resources Window**.

© *Eastwood Harris*

3.2.12   Personal Identifiable Information Tab

The **Personal Information** tab was new to P6 Version 18 and displays your personally identifiable information (PII) that the administrator enters when the user was created.

This information is available from the **Admin** menu when a PPM database is opened. The **Admin** menu has been removed when the user opens an EPPM database and this tab enables users to see their Personal Information without requiring access to the Web client.

This personal information may be exported with project data and this function was introduced at the same time as **Admin Preferences**, **Consent Notice** functions:

Personal Information
Login Name
harrispe
Personal Name
Paul E Harris
Phone
+61 (0)41 118 7701
Email
harrispe@eh.com.au
Associated Resource
Paul E Harris

Forget User Acceptance

Your personal information is updated by User Admin. You may get in touch with User Admin to update it.

3.3 Default Toolbar issues

There are two issues with the default Toolbar settings:

❖ Firstly, the **No Bottom Layout** button is missing from the toolbar, making it difficult to hide the bottom pane.

❖ **NOTE:** This button should be added by clicking on the down arrow as displayed in the picture below:

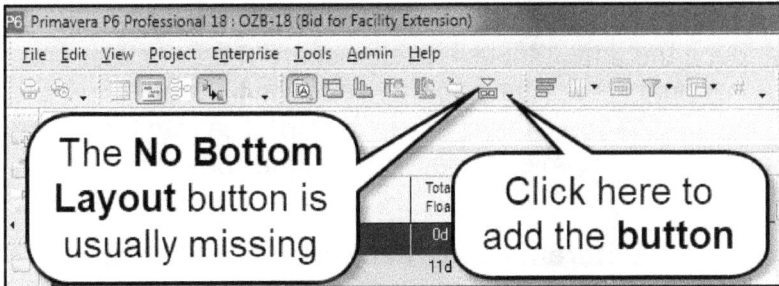

The **No Bottom Layout** button is usually missing

Click here to add the **button**

❖ Secondly, full menus are not available and so only some of the commands are displayed when you click on a menu.

❖ **NOTE:** You should select the **View, Toolbars...,** **Customize, Options** tab and uncheck the **Show full menus after a short delay** and the **Menus show recently used commands first** options in order to ensure full menus are always displayed:

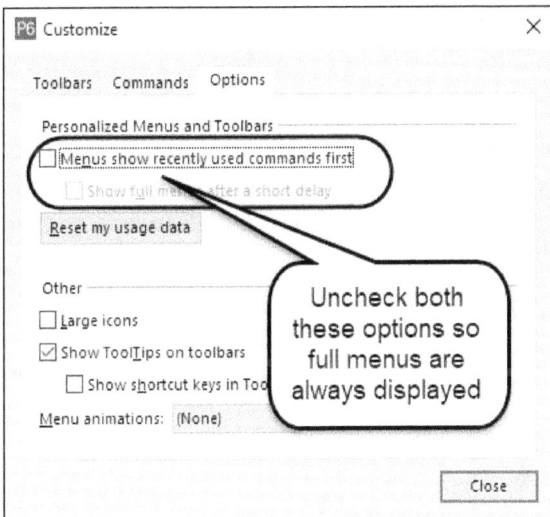

Uncheck both these options so full menus are always displayed

3.4   Set Language

A number of menu languages are available.

Version 15.2 moved the selection of the menu language from the **Tools** menu to the Login screen.

3.5   Portfolios

It is important for users to know that Portfolios not only reduce the number of projects that are displayed but also reduce the time it takes for P6 to open. Thus, if a user has many projects available to them and the software takes a long time to open, then the use of Portfolios with the regularly used projects may assist in the software opening more quickly.

3.5.1   Difference between User Baselines and Project Baselines

User Baselines are not **Project Baselines**.

❖ When a second user opens a project, which has a **Primary User Baseline** set by the first user, then this baseline will not be assigned to the second user.

❖ When the same layout is used to display the project by a second user, the **<Current Project> Baseline**, which displays the **Planned Dates**, will be displayed as the **Primary User Baseline**.

❖ **NOTE:** Therefore, two users opening the same project and using the same Layout may display different data.

3.6   Who has been messing with my project?

If you think someone has been messing with your project, then consider using the **Audit Trail Columns** to track down the culprit.

Primavera Version 5.0 introduced four basic audit trail columns that may be displayed in the Activities Window, which display the date and user who added the activity and by whom and when it was modified:

❖ **Added By** – the user who added the activity,

❖ **Added Date** – the date the activity was added,

❖ **Last Modified By** – the user who last modified the activity, and

❖ **Last Modified Date** – the date the activity was last modified.

Primavera Version 6.0 introduced two new resource assignment fields available in the **Activities Window, Activity Details, Resources** tab:

❖ **Assigned by**, and

❖ **Assigned Date**.

3.7   Closing Down

The closing down options are:

❖ Select **File, Close All** or **Ctrl+W** to close all Projects.

❖ Select **File, Exit** or click the ⊠ icon in the top right side of the Primavera window to shut down all projects and close Primavera.

❖ If you close down the system leaving one or more projects open, then these projects will be open the next time you log in.

❖ Go to the **Edit, User Preferences..., Application** tab and set the **Application Startup Window** to **Activities** so the software will open with your last project Activities window displayed.

4   CREATING A NEW PROJECT

4.1   New Project Wizard

When a user creates a new project using the **File, New** command, it runs the **New Project Wizard**.

❖ The project will be assigned the P6 defaults and

❖ Some of these should be changed immediately.

4.1.1   Date and Time

Before creating a project, a user must ensure that the date and time are being displayed.

4.1.2   Close all other projects

If you create a project with another open, you will then have two projects open, so it is recommended that you close down all other project before creating a new one.

4.1.3   EPS Node

The user should ensure that a suitable EPS Node is available for the new project.

4.1.4   Project ID

An organization should have a procedure, so all projects are easily identified by the Project ID. It should allow for parameters such as:

❖ Type of project, Bid, What If etc.

❖ Client

❖ Version

The author recommends:

❖ The EPS node should have the organization's project number as the EPS ID.

❖ Each Project should have the Project ID plus an extension identifying the type of project and version, or Data Date or both version and Data Date.

❖ Each time a project is updated it should be copied for reports etc. The scheduler should continue working on the same project file, NOT a copied file. This is important, especially if the scheduler is using WBS filters, as these only work on the one project and will not work on a copied project, as the WBS includes the P6 Project ID.

4.1.5   Project Must Finish By Date

This is an optional date and when set, it is used to calculate Total Float.

❖ The author recommends that this date is not set because the Critical Path disappears when the date is set later than the last activity finish date and all activities will have Total Float.

❖ P6 will set the default project **Must Finish By** date at midnight or at the start of the day, effectively giving one less day than anticipated. So, ensure you always display the time with the date.

❖ If you anticipate using this function, then ensure you have the time displayed so you can change the time setting to the end of the day.

4.2   Other Methods of Creating a Project

The other methods of creating a project are:

❖ **Copy** and **Paste** a project, groups of projects or even an EPS Node. The P6 process of copying and pasting projects is very powerful, allowing a lot of control over what is and what is not copied. P6 automatically adds numbers to the Project IDs to provide a unique Project ID, which may then be edited.

❖ Import a project created in another P6 database that has been exported in P6 **XER** or **P6 XML** format. Often, importing a project from another database brings in a lot of unwanted data from the other database and administrators usually check and clean up the file before importing it, with a third part tool. There are many third-party tools listed at **www.primavera.com.au**.

❖ Create a project and import data using **Excel**. Usually an Excel spreadsheet is exported, populated in Excel with project data and imported.

❖ Import a project created in Microsoft Project. The file should be saved in **Microsoft Project XML** format and imported into P6 using the P6 Import function. This process has a number of issues and the user needs to be careful when importing data from a Microsoft project to ensure that they are not going to overwrite existing data and check the imported P6 file thoroughly for variances.

NOTE: the **Microsoft Project XML** format is the same language as the **P6 XML** file, but the data is formatted differently. It is recommended that you put P6 or MSP in the file name when you are using both P6 and Microsoft Project XML files.

4.3 Setting Project Defaults and Other Data in the Projects Window Details Pane

Consider changing the following fields after running the **New Project Wizard** or creating a project any other way, when you are creating an un-resourced project.

4.3.1 General Tab, Status

The **Status** field is useful for sorting and filtering projects.

❖ Often you will end up with multiple schedules for one project and this field allows you to easily isolate the current project and filter out old versions of your schedule.

❖ Corporate project procedures should specify the use of the **Status** field.

4.3.2 Dates Tab, Scheduled Dates

❖ The **Planned Start** is the date that no activity will be scheduled to start before. E.g. the Project Start Date.

❖ The **Must Finish By** date is an optional date. When this date is entered it is used to calculate the **Late Finish** of activities, thus all **Total Float** will be calculated to this date.

NOTE: It is recommended that under normal circumstances this should never be set and left blank.

© *Eastwood Harris*

These are short explanations of the following fields:

❖ The **Finish date** is a calculated date and is the date of the completion of the last activity.

❖ The **Data Date** is used when updating a project. Unlike Microsoft Project, in P6 all incomplete work is scheduled to take place after this date.

❖ The **Actual Start** date is inherited from the earliest started activity.

❖ The **Actual Finish** date is inherited from the latest completed activity when all activities are complete.

4.3.3 Dates Tab, Anticipated Dates

The **Anticipated** dates are useful for corporate long-term planning when the details of a project are unknown, and they allow a project bar to be displayed in the Projects window when there are no activities.

❖ The **Anticipated Start** and **Anticipated Finish** dates may be assigned before a WBS structure and Activities have been created.

❖ The **Start** and **Finish** dates columns and bars at the EPS level adopt the **Anticipated** dates when there are no activities. This allows long term planning of projects without adding any activities and is very useful in a company planning long term maintenance work:

Project ID	Project Name	Total Activities	c 27	Jan 03	Jan 10
⊟◆ **SDP**	**Shut Down Planning**	0			
SD-01	Shutdown 01	0			
SD-02	Shutdown 02	0			
SD-03	Shutdown 03	0			

❖ After Activities have been created, the **Anticipated** dates may remain as a historical record only and are not displayed or inherited anywhere else.

4.3.4 Defaults Tab, Defaults for New Activities

Duration Type

❖ The **Duration Type** options ONLY affect how the schedule calculates after one or more resources is assigned to an Activity.

❖ The following options are available:

  ➢ **Fixed Units**
  ➢ **Fixed Duration & Units/Time**
  ➢ **Fixed Units/Time**
  ➢ **Fixed Duration & Units**

❖ If you do not plan to add resources to Activities, then you do not need to assign a **Duration Type** and it may be left as the default.

❖ If you do plan to add resources, the author recommends one of the following options:

  ➢ **Fixed Duration & Units/Time** if you wish the crew size to remain the same as you increase of decrease the activity duration, but the estimate to complete will change

  ➢ **Fixed Duration & Units** if you wish the estimate to complete to remain the same as you increase of decrease the activity duration, but the crew size will change.

  ➢ **NOTE:** The author recommends NOT to use **Fixed Units** and **Fixed Units/Time** because they result in the duration changing each time a resource is added or removed, which is normally undesirable.

Percent Complete Type

Each new activity **Percent Complete Type** is set to the **Default Percent Complete** and may be changed at any time in the Activity Details, **General** tab:

❖ **Duration % Complete** – This field is calculated from the proportion of the **Original Duration** and the **Remaining Duration** and they are linked, and a change to one value will change the other. When the **Remaining Duration** is set to greater that the **Original Duration**, this percent complete is always zero. This is similar to the way Microsoft Project % Complete calculates.

❖ **Physical % Complete** – This field enables the user to enter the percent complete of an activity and this value is independent of the activity durations, cost and units. **NOTE:** This is the authors recommended setting and must be selected if you wish to use **Steps**.

❖ **Units % Complete** – This is where the percent complete is calculated from the resources Actual and Remaining Units. A change to one value will change the other and when more than one resource is assigned, then all the Actual Units for all resources will be changed proportionally. This is similar to the Microsoft Project % Work Complete.

❖ The **Activity % Complete** field is linked to the **% Complete Type** field assigned to an activity in the General tab of the **Details** form in the **Activities Window** or the **% Complete Type** column:

❖ The **Activity % Complete** is also linked the **% Complete Bar** and this value is represented on the **% Complete Bar.**

Percent Complete Type	Orig Dur	Rem Dur	Activity % Complete	Duration % Complete	Physical % Complete	Units % Complete	Actual Labor Units	At Completion Labor Units	Jan 05	Jan 12
Duration	10d	6d	40%	40%	12%	0%	0	0		
Physical	10d	6d	12%	40%	12%	0%	0	0		
Units	10d	6d	60%	40%	12%	60%	12	20		

Cost Account

Left blank, as P6 will assign the same Cost Account to all new resources, which most projects do not require.

Calendar

This is where the new activity default calendar is set.

❖ A new **Project Calendar** should be created first and set here as the **Activity Default Calendar**.

❖ **NOTE:** The author recommends that a **Project Calendar** should normally be used for all activity calendars because **Global Calendars** may be more easily changed by someone else, without your knowledge, and your schedule will calculate differently.

4.3.5   Calculations tab, Activities

If you are creating an un-resourced schedule then the defaults set here should normally not be adjusted, except if you intend to use **Steps**, then you must use:

❖ **Physical % Complete** and

❖ Check the **Activity percent complete based on activity steps** check box.

4.3.6   Calculations tab, Resources Assignments

The defaults set here should normally not be adjusted when you are creating an un-resourced schedule.

4.3.7   Codes tab

If your organization is using Project Codes to Group and Sort projects, then you should assign them as per your organization's procedure.

4.3.8   Notebook tab

You should add any notes about your project here using the appropriate **Notebook Topics**.

4.3.9   Resources tab

The defaults set here should normally not be adjusted when you are creating an un-resourced schedule.

❖ **Specify the default Rate Type for new assignments** – Resources may be assigned up to five rates and:

➢ This is set to the default rate when resources are assigned and

➢ May be changed for each assignment individually in the **Activity Details, Resources** tab after the resource has been assigned.

❖ **NEVER EVER** uncheck **Drive activity dates by default**, otherwise you may end up with resources scheduled outside an activity duration and after the activity has finished.

❖ **Resources can be assigned to the same activity more than once** – This is useful in circumstances when you require to assign a crane on the first and last day of an activity. This is achieved by using the **Resource Lag** and **Resource Duration**,

4.3.10   Settings tab

Summarized Data

See the **Summarized Data** section for information on **Summarization**.

Project Settings

Character for separating code fields for the WBS tree - There may be a requirement to use a specific WBS Node separator for contractual or system requirements:

❖ The default **WBS Node Separator** is assigned in the **Admin, Admin Preferences..., General** tab.

❖ Each individual project **WBS Node separator** is defined in the **Projects Window, Project Details** form, **Settings** tab and overrides the default set in the **Admin Preferences** form, **General** tab.

❖ There may be a requirement to have specific characters for a WBS Code either:

➢ To meet client reporting requirements, or

➢ Allow the importation of data from another system.

4.4 Tools, Schedule Options

The **Schedule Options** are edited by selecting **Tools, Schedule, Options**.

❖ The default settings are normally good, but you should consider a changing some when the situation arises, the rest are best left as default:

In P6 Version 20 and later, select the project **Scheduling Options** which will be assigned permanently to all open project and used to calculate all open projects

Select **Make open-ended activities critical** when multiple critical paths are required

Select **Longest Path** when using multiple calendars are assigned to activities

Default Project

❖ When more than one project has been opened then it is important to understand how the **Default Project** and **Use scheduling options from** functions operate.

❖ **Use scheduling options from** was introduced in P6 Version 20 and overrides the project selected in **Project, Default Project.**

❖ In summary:

  ➢ The selected project **Scheduling Options** or Default Project in Version 19 and earlier are used to schedule all the opened projects, but

  ➢ More importantly all the project's **Scheduling Options** of all opened projects are changed permanently to the **Default Project Scheduling Options** or Default Project in Version 19 and earlier and they will not calculate the same way again.

❖ To prevent calculation issues, it is best to assign all projects opened together the same **Scheduling Options.**

Make open-ended activities critical

❖ You should select **Make open-ended activities critical** when multiple critical paths are required, and

Longest Path

❖ Select **Longest Path** when multiple calendars are being used.

4.5 Author's recommended new project setup

Step	Suggested Settings
Set the **Units/Time** format by selecting **Edit**, **User Preferences...** to open the **User Preferences** form and select the **Time Units** tab.	There is a choice of **percentage (50%) or units/duration (4h/d)**. This should be set on personal preference. The author prefers **(4h/d)** as this reduces typing. The **User Preferences**, **Time Units** setting affects how these are displayed and 16h/d or 2d/d or 200% is two people.
Set the **Resource Assignments** option by selecting **Edit**, **User Preferences...** to open the **User Preferences** form and select the **Calculations** tab.	It is suggested that the **Preserve the Units, Duration, and Units/Time for existing assignments** is selected. With this option, as Resources are added or deleted, the total number of hours assigned to an Activity increases or decreases. Each Resource's hours are calculated independently. The options under **Assignment Staffing** need to be carefully considered and understood so that when Resources are assigned to Roles and resource assignments are changed, that the user understands which Unit Rate and which Unit Cost will remain against the activity.
In the **Project Window**, **Defaults** tab set the default **Activity Type**.	It is suggested that **Task Dependent** is used, as with this option Resource calendars are not used, making the schedule simpler.

Step	Suggested Settings
In the **Project Window**, **Defaults** tab set the default **Duration Type**.	It is suggested that **Fixed Duration & Units** is used as a default. With this option the Activity Duration does not change when resource assignments are altered, and when an Activity Duration is changed, the Units do not change, so your estimate of hours and costs will not change. When you wish the crew size to remain the same when the activity duration is changed then you should select **Fixed Duration & Units/Time**. The cost and units will change proportionally.
In the **Project Window**, **Defaults** tab set the default **Percent Complete Type**.	The author prefers to use **Physical** as this enables the **Activity Percent Complete** to be independent of the **Activity Durations**, and **Steps** may be used when updating a project.
In the **Project Window**, **Resource** tab set the default **Resource Assignment Defaults**.	Unless multiple Rates are being used then **Price/Unit** should be selected. Check **Drive activity dates by default**.

5 CALENDAR ISSUES

5.1 Non-Work time display

The **Database Default Calendar** is selected in the **Enterprise**, **Calendars...** form and is used to display the Nonworking times for all users in all layouts and for all projects in a database.

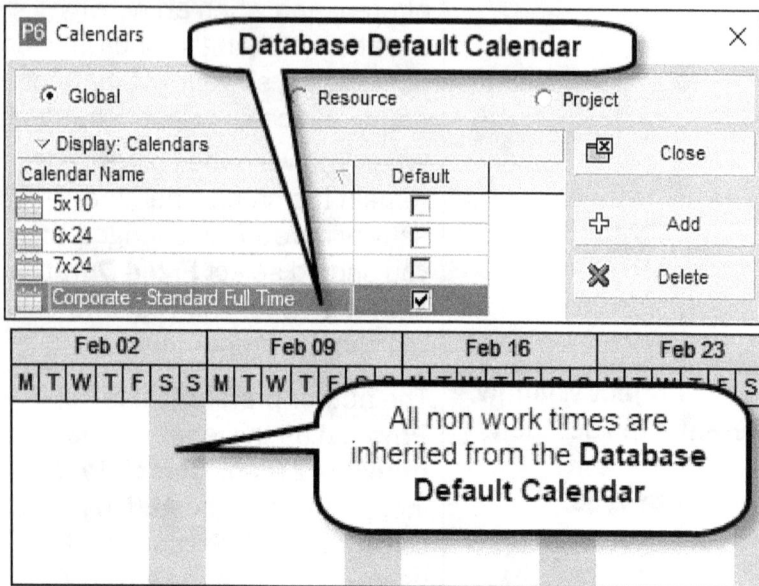

It is not possible for users to display different nonwork periods for different projects or views, as in most other scheduling software packages, without affecting all other projects in a database.

For example, Microsoft Project may display a different calendar non work time in each different View and Elecosoft (Asta) Powerproject may display each activity calendar as shading behind each activity bar.

This may become an issue with projects that have different work periods and rosters. This issue may be solved by creating another database with a different **Database Default Calendar**.

5.2 Copying Project Calendars

Project calendars many not be copied, but may be moved to become a Global Calendar:

To copy a calendar from one project to another:

❖ Move the Project Calendar to be a Global Calendar,

❖ Create a new Project Calendar in both projects by copying the new Global Calendar.

❖ You will have to assign the newly created calendar to any existing activities as they will now be assigned the original calendar, that will now be a Global calendar.

❖ **NOTE:** If you delete the Global calendar you will be offered the option of selecting a replacement calendar.

5.3 Inherit Holidays and Exceptions from a Global Calendar

When creating a new Project or Resource calendar, a Global Calendar may be selected from the drop-down box and this function will link the calendar holidays from the selected Global Calendar into the displayed calendar.

The Global and the new Project or Resource calendars will remain linked and a change to a Global calendar holiday will be reflected in a calendar with Inherited Holidays.

NOTE: It is suggested that this option **NEVER** be used so each calendar is created as standalone without inheriting holidays from another calendar, and therefore will not change if another calendar has holidays changed.

5.4 Calendars for Calculating Project, WBS and Other Summary Durations

In Primavera Version 7 and later, the summary durations of bands projects in the **Projects Window** are calculated based on the **Database Default Calendar** from the first activity Start to the last activity Finish, from all the project in that band.

The summary duration of WBS bands and other bands created by Grouping activities by **User Defined Fields** or Activity Codes are calculated by:

❖ When all the activities in a band share the same calendar, then the summary duration is calculated on the calendar of the activities in the band and the calendar name is displayed in the Calendar column.

❖ When the calendars for the activities are different, the summary duration is calculated on the Project Default calendar and the calendar field is blank.

The Project Default in the picture below has the calendar set as the 8hr/d & 5d/w and the Project Default calendar is used to calculate the summary duration for WBS Nodes, Projects when the calendars are different:

	Calendar	Original Duration	Dec 06							Dec 13				
			n	Mon	Tue	Wed	Thr	Fri	Sat	Sun	Mon	Tue	Wed	Thr	Fri
Calendars Durations		10d												
Calendars = Project Default	8hr/d & 5d/w	10d												
A1030	8hr/d & 5d/w	10d												
A1040	8hr/d & 5d/w	7d												
A1050	8hr/d & 5d/w	5d												
Calendars all different		10d												
A1060	7 x 24hr. Days	12d												
A1070	8hr/d & 7d/w	12d												
A1080	8hr/d & 5d/w	10d												
Calendars NOT = Project Def	7 x 24hr. Days	12d												
A1090	7 x 24hr. Days	12d												
A1100	7 x 24hr. Days	5d												
A1110	7 x 24hr. Days	8d												

5.5   Resource Dependent Activity Start Date and Time

When an activity is made **Resource Dependent**, unlike in some other software, it still acknowledges the Activity Calendar for calculating the start of the resource work.

Activity Type	Calendar	Resources	Aug 03						
			Fri	S	S	M	T	W	T	Fri
Resource Dependent	5 x 8 No Hols								
Resource Dependent	5 x 8 No Hols	24 h/d 7d/w							
Resource Dependent	7 x 24 No Hols	24 h/d 7d/w							

This activity is scheduled to start on the Data Date ONLY when both the resource and activity are assigned a 24x7 calendar

5.6 Project or Global calendars

The author recommends the use of Project Calendars in all project environments:

❖ These may only be edited when the project is open.

❖ When Global Calendars are assigned to activities, then these calendars may be edited without the project scheduler's knowledge and the schedule will calculate differently.

❖ Maintenance programs are often run with Global Calendars, but these need to be strictly controlled.

5.7 Elapsed Duration Activities

Activities like "concrete curing" and "process testing" that take place 24 hours per day may be modeled using a 24 hour per day calendar. The disadvantage of this option is that if the predecessor ends at 17:00 then a 5-day curing activity will start and finish at 17:00 and will have float when the successor starts the following morning. To prevent this, I recommend putting these types if activities on calendars with the same hours per day as the other work, but on a 7 day/week basis without weekends or non-workdays:

Activity Name	Calendar	Orig Dur	Start	Finish	Total Float	Jan 09 · Jan 16 · Jan 23
Elapsed Duartions						
All Activities on a 8hr/day 5 days/week						
Pour	8hr/d 5d/w	5d	09-Jan-23 08	13-Jan-23 16	0d	
Cure	8hr/d 5d/w	5d	16-Jan-23 08	20-Jan-23 16	0d	
Strip	8hr/d 5d/w	5d	23-Jan-23 08	27-Jan-23 16	0d	
Curing on 8 hour per day 7days/week						
Pour	8hr/d 5d/w	5d	09-Jan-23 08	13-Jan-23 16	0d	
Cure	8hr/d 7d/w	5d	14-Jan-23 08	18-Jan-23 16	0d	
Strip	8hr/d 5d/w	5d	19-Jan-23 08	25-Jan-23 16	0d	
Curing on 24 hour per day 7days/week						
Pour	8hr/d 5d/w	5d	09-Jan-23 08	13-Jan-23 16	0d	
Cure	24hr/d 7d/w	5d	13-Jan-23 16	18-Jan-23 16	0d 15h	
Strip	8hr/d 5d/w	5d	19-Jan-23 08	25-Jan-23 16	0d	

NOTE: If the curing activity ends on Saturday, then it will normally still have Total Float when the successor may not start till the following Monday.

5.8   Tips for Mixed Calendar Schedules

When a project has mixed calendars, say an 8- and 10-hour per day, then a change of calendar from a predecessor on an 8-hour per day calendar to a successor on a 10-hour calendar, the successor activity may have one hour of work on the same day as the predecessor and span 2 days. This situation leads to interesting Float calculations and confusion to schedulers.

Calendar	Original Duration	Mon Apr 02																Tue Apr 03																
		8	9	10	11	12	13	14	15	16	17	18	19	20	21	22	23	0	1	2	3	4	5	6	7	8	9	10	11	12	13	14	15	16
5d/w 8:00-16:00	1d																																	
5d/w 7:00-17:00	1d																																	

Primavera P6 does not have a "Start on a New Day" function found in other products, such as Elecosoft (Asta) Powerproject, but which in itself brings on a new set of calculation issues.

Techniques that may be considered to ensure one-day activities span one day and two-day activities span two days, etc. are:

❖ Apply an appropriate lag to the relationship, not recommended and many contracts disallow this type of relationship, or

❖ Ensure all calendars have the same finish time but different start times thus giving the activities a different number of hours per day. Then as long as the activity durations are whole days, they will be scheduled to start at the start of the day according to their calendar and will end at the same time at the end of the day, or

❖ When the Start and Finish Times are not an important scheduling consideration, then assign all the calendars the same Start and Finish time, but adjust the duration of the lunch break, so the days have the desired number of hours.

© *Eastwood Harris*

The picture below shows how 8-hour a day and 12-hour a day calendars in the same project may be set up to ensure all activities start and finish on the same day:

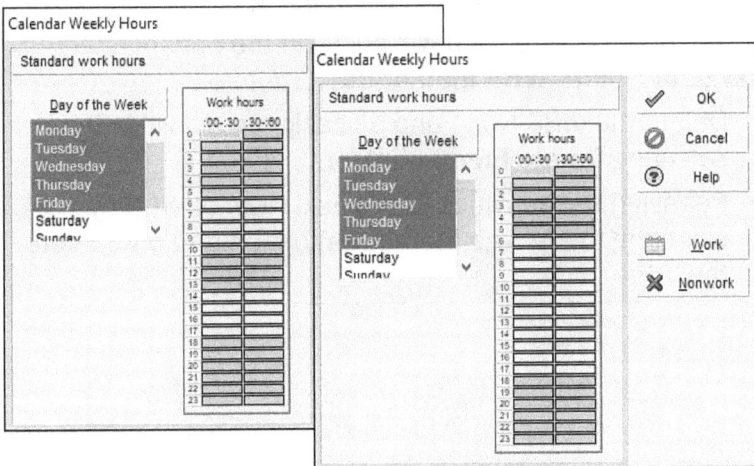

6 CREATING A WBS

6.1 Merge or Delete a WBS Node

When a WBS Node has been assigned activities, and you delete the WBS Node you will be given the option in the **Merge or Delete WBS Element(s)** form to either:

❖ Delete the WBS Node and all assigned activities by selecting **Delete Element(s), or**

❖ Reassign the activities to the next level of WBS by selecting the **Merge Element(s)** option and only delete the WBS Node:

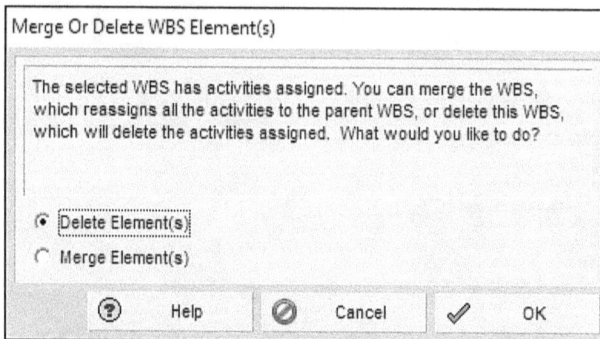

Merge Or Delete WBS Element(s)

The selected WBS has activities assigned. You can merge the WBS, which reassigns all the activities to the parent WBS, or delete this WBS, which will delete the activities assigned. What would you like to do?

⊙ Delete Element(s)

◯ Merge Element(s)

⊘ Help ⊘ Cancel ✓ OK

6.2 WBS Node Separator

There may be a requirement to use a specific WBS Node separator for contractual or system requirements for import or export of data:

❖ The **Default WBS Node Separator** is assigned in the **Admin, Admin Preferences..., General** tab.

❖ Each individual project WBS Node separator is defined in the **Projects Window, Project Details** form, **Settings** tab and overrides the default set in the **Admin Preferences** form, **General** tab.

6.3 WBS Categories

WBS Nodes may be assigned categories, which enable WBS Nodes to be grouped and sorted under the WBS Categories.

WBS Categories are to WBS Nodes as Activity Codes are to Activities. In earlier P6 versions these were referred to as Project Phases.

One use of a WBS Category could be, for example, to tag the WBS Nodes with the phases such as Design, Procure and Install and then the Activities may be grouped by WBS Category and then by WBS or some other Activity Code.

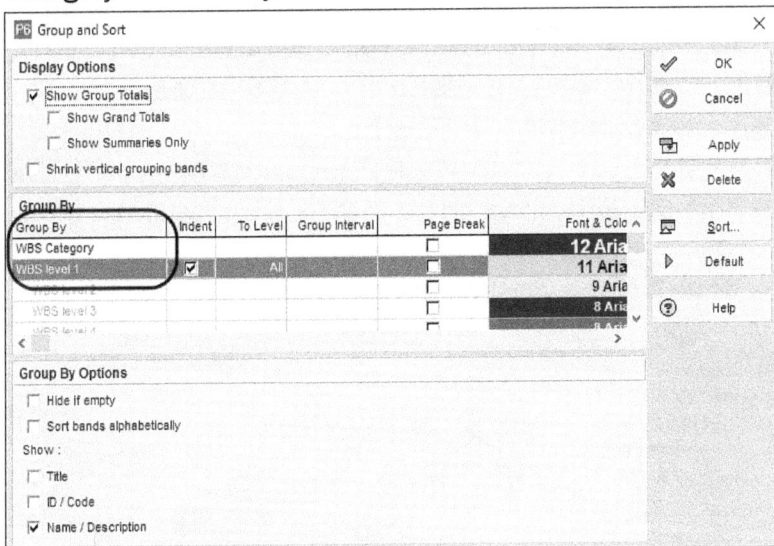

WBS Categories are created using:

❖ In the Professional Client in the **Admin**, **Admin Categories...**, **WBS Categories** tab, and

❖ In the **Web Client**, **Administer**, **Enterprise Data**, **Projects WBS Categories**.

6.4 Why is the WBS not displayed correctly?

The WBS may be displayed incorrectly after you have added some WBS Nodes and returned to the Activities window. For example, new WBS Nodes may be displayed hierarchically when they should not be; if so just press F5 to refresh the screen. The pictures below show before and after pressing F5:

Activity ID	Activity Name		Activity ID	Activity Name
SD-01 WBS Node Example			SD-01 WBS Node Example	
SD-01.1 Design			SD-01.1 Design	
SD-01.2 Procure			SD-01.2 Procure	
SD-01.3 Install			SD-01.3 Install	

6.5 Why can I not see the WBS in the Activities Window?

You will not see WBS bands in the Activities windows if:

❖ The **Group By** must be set to WBS level 1 and To Level set to All, and/or

❖ **Hide if empty** must not be checked.

© *Eastwood Harris*

7   ADDING ACTIVITIES

7.1   New Activity Defaults

After creating a new project and before adding activities, it is important to set the defaults such as the Activity ID Auto-numbering defaults, Duration Type and Calendars.

By setting them correctly before adding activities you will save a significant amount of time, because you will not have to change a number of attributes against all activities at a later date.

These defaults are set in the **Defaults** tab of the **Project Details** form:

Many of these defaults were introduced earlier but we will go through them in detail here.

7.2   Duration Type

The **Duration Type** options only affect the schedule calculations after one or more resources are assigned to an Activity.

The following options are available:

❖ **Fixed Units**
❖ **Fixed Duration & Units/Time**
❖ **Fixed Units/Time**
❖ **Fixed Duration & Units**

If you do not plan to add resources to Activities, then you do not need to assign a **Duration Type** and it may be left as the default.

If you wish to assign resources, then one of the two following options are recommended:

Fixed Duration & Units

With this option:

❖ The activity duration will not change when resources are assigned or removed from an activity,

❖ A change in duration will not change the number of hours and costs assigned to an activity, and

❖ The crew size will change as the activity duration is changed.

❖ If one person is assigned to an activity for 8 hours per day and the activity is doubled in duration, there will now be one person working on the activity for 4 hours per day and the activity will require the same number of hours to complete.

Fixed Duration & Units/Time

❖ The activity duration will not change when resources are assigned or removed from an activity, and

❖ A change in duration will keep the crew size the same, and

❖ The hours and cost will change as the duration is changed.

❖ If one person is assigned to an activity for 8 hours per day and the activity is doubled in duration, there will still be one person working on the activity for 8 hours per day and the activity will require double the number of work hours to complete.

7.3 Percent Complete Type

The default for new projects is **Duration % Complete**.

❖ The default for new activities is set in the **Project Details** window, **Defaults** tab, **Percent Complete Type** drop down box.

❖ This option may be set for each activity individually in the **Activities** window **Details**, **General** tab.

NOTE: The author recommends **Physical % Complete** and this should be set in the **Defaults** tab of the **Project Details** windows, **Defaults** tab before any activities are added.

❖ This option unlinks the **Remaining Duration** and **Activity % Complete** and allows them to be entered separately;

❖ Thus, the **Remaining Duration** does not change when the **Activity % Complete** is changed.

❖ This in turn allows the maintenance of **Remaining Durations** in whole days.

❖ Also, **Physical % Complete** is required when using **Steps** to measure progress.

7.4 Cost Account

The drawback of the P6 **Cost Account** function is that all resources are assigned the same value, which is normally undesirable.

Cost accounts are normally used to group costs by parameters required by the accounts department, such as permanent or temporary materials, and used for depreciation etc.

❖ It is more normal to assign a Cost Account to a resource and when a resource is assigned to an activity, it automatically has the desired Cost Account.

❖ If you wish a more traditional Cost Account set-up then you may wish to consider using a Resource Code to represent a Cost Account.

7.5 Calendar

As stated earlier, this should normally be set as a **Project Calendar**.

NOTE: The author recommends that a **Project Calendar** should normally be used for all activity calendars, because **Global Calendars** may be changed by someone else, without your knowledge, and your schedule will calculate differently.

7.6 Copying Activities from other Programs

Activity data may **NOT** be copied from or updated from other programs (such as Excel) by cutting and pasting.

7.7 Copying Activities in P6

Activities may be copied from another project when both projects are open at the same time, or copied from within the same project using the normal Windows commands **Copy** and **Paste**, by using the menu commands **Edit, Copy** and **Edit, Paste**, or by using **Ctrl+C** and **Ctrl+V**.

The **Copy Activity Options** form will be displayed. These options are self-explanatory.

7.8 Renumbering Activity IDs

There is a new function in P6 Version 7 allowing the renumbering of activities. To use this function:

❖ Select the activities that are to be renumbered,

❖ Select from the menu **Edit**, **Renumber Activity IDs** or right-click in the columns area and select **Renumber Activity IDs**,

❖ This opens the **Renumber Activity IDs** form, allowing renumbering of the activity IDs.

❖ You must remember to renumber the Baseline projects by checking the **Renumber selected activities in baselines**.

NOTE: The P6 **Renumbering** function does not renumber the project Activity IDs in the order displayed on the screen unless you go into **Group and Sort** form, select **Sort** and

---

delete all the Sorts in this form. Then P6 will renumber all the activities from top to bottom.

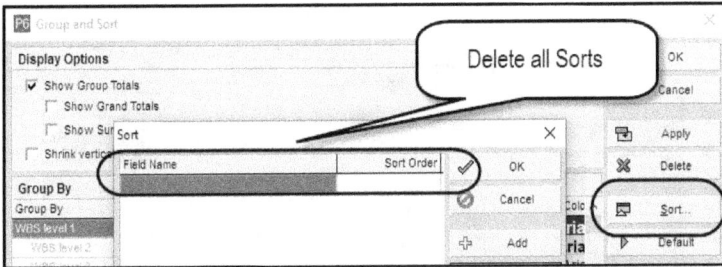

7.9 Finding the Bars in the Gantt Chart

At times you will find there are no bars displayed in the Gantt Chart, because the Timescale has scrolled too far into the past or future.

Double-click in the Gantt Chart in line with an activity and the Timescale will scroll to display the activity bar.

7.10 Assigning Calendars to Activities

Activities often require a different calendar from the default **Project Calendar** that is assigned in the **Project Information** form. Primavera enables each activity to be assigned a unique calendar.

An **Activity Calendar** may be assigned from:

❖ The **General** tab of the **Bottom Layout** or

❖ By displaying the **Calendar** column.

7.11 Assigning Activities to a WBS Node

Activities are assigned to a WBS Node from the Activities Window. They may be assigned a WBS Node using the following methods:

❖ A new activity will inherit the WBS Node that is highlighted when an activity is created.

❖ A new activity will inherit the WBS Node of a selected existing activity when the project is organized by WBS Nodes and an activity is created.

❖ Select the activity and click the WBS box in the **General** tab in the lower window. This will open the **Select WBS** form where you may assign the WBS Node.

❖ Select one or more activities and move the mouse to the left of the activity description and the mouse will change into the shape displayed in the following picture. You may then drag the activities to another WBS Node:

Activity ID	Activity Name	Original Duration
Bid for Facility Extension		31d
Tech...	Be sure the mouse pointer changes to this shape before dragging	13d
0Z1		0d
0...	...nts	4d
0Z1020	Create Technical Specification	5d
0Z1030	Identify Supplier Components	2d
0Z1040	Validate Technical Specification	2d
Delivery Plan		14d
0Z1050	Document Delivery Methodology	4d

❖ Insert the WBS column and use the **Fill Down** command

7.12 Adding Splits to an Activity

If you have created a activity and you wish to split it then you have the following options:

❖ If the activity has started and therefore has an Actual Start you may create one split using the **Suspend and Resume** function.

❖ Multiple splits may be modeled with a specific calendar and the calendar bar Necked to show the splits but scheduling after changes to the predecessors this may not give the desired results.

❖ You may break the original activity down into multiple activities. To keep the original baseline you may make the original activity a LOE and use it to span over multiple new activities. Now the original baseline for the LOE activity will be valid but will not display unless you format the baseline bars to show Baseline Bars for **All Activities** as per the picture below and not as **Normal** which is the default:

Display	Name	Timescale	Filter	Preview
☑	Primary Baseline	Primary Baseline Bar	All Activities	▭
☐	Primary Baseline	Primary Baseline Bar	Milestone	▽ ▽
☐	Primary Baseline	Primary Baseline Bar	Summary	▭

7.13 Reordering or Sorting Activities

The sort order of activities within a band is set by an order from one or more columns and you may not drag activities up or down the schedule in the same way as other products.

To reorder activities either:

❖ Click on a column header, the arrow in the header will indicate the sort order, or

❖ Use the **Sort** option in the **Group and Sort** form.

7.14 Undo

The **Undo** function that operates on Resources, Resource Assignments, and Activities Windows, but no **Redo** function.

There are many functions that will erase the Undo memory such as scheduling, summarizing, importing, opening a project, opening Code forms, opening **User and Admin Preferences** and closing the application.

7.15 Spell Check

To spell check a project, open the **Spell Check** form:

❖ Select **Edit**, **Spell Check**, or

❖ Hit the **F7** key.

NOTE: You must select a cell in the column you wish to spell check on. Do not click on the column header to spell check as this will resort the activities on that column.

7.16 Updating Parameters in a Large Number or All Activities

If you wish to change the parameters of some or all activities in a project, such as the Percent Complete Type or Duration type the are two ways to achieve this. Simply:

❖ **Edit**, **Fill Down** (**Ctrl+E**), or

❖ Set up a **Global Change**.

8   FORMATTING

The formatting of all P6 windows is very similar. Thus, Grouping and Sorting projects in the Project window is very similar to Grouping and Sorting activities in the Activities window.

8.1   Activity Window Bar Formatting

There are many issues with all the formatting of the bars in all of the Oracle Primavera P6 Activity Layouts supplied with the software. Also **NEVER** select **Default** in the **Bars** form as this will apply some bad bar formatting to your layout.

The author has placed a layout on www.primavera.com.au and www.eh.com.au under **Technical Papers**, titled **www.primavera.com.au_layout** that has fixed the bar formatting issues discussed below.

It is suggested that downloading this layout will save users a significant amount of formatting on their existing layouts.

The bar formatting from this Layout may be copied to any other Layout in the **Bars** form, **Copy from...** function.

The issues that are prevalent in some or all of the Oracle Primavera Layouts are:

❖ There is Total Float displayed on complete activities,

❖ Relationships are displayed on the Baseline Bars,

❖ The Baseline bar formatting for all Baseline bars are all the same size, color and in the same position, thus it is difficult to see which baseline is being displayed,

❖ Three Baseline Milestones will not be displayed,

❖ Bar Text is assigned to a number of bars. Bar text should be assigned to a blank bar, so it may be simply hidden or displayed.

❖ LOE bars are not displayed, so they are hidden when created.

The full explanation of the issues and fixes made in the **www.primavera.com.au_layout** are documented in the authors other P6 books.

You have several choices for sorting out the bar formatting:

❖ Delete the Oracle Layouts and use the **www.primavera.com.au_layout** as a basis for all your layouts, or

❖ Copy the **www.primavera.com.au_layout** bar formatting to all other Layouts including the Oracle Layouts, using the **Bars** form **Copy From..** function or

❖ Edit the layouts you are using.

8.2   Creating a Summary Bar

It is not obvious how to create a Summary bar:

❖ When creating a new Summary bar, you will not be able to select **Summary** from the filter drop down box,

❖ You must check the **Bar Settings** tab, **Show bar for grouping bands** to create a Summary Bar:

Display	Name	Timescale	User Start Date	User Finish Date	Filter	Preview
☐	Activity Name on	Current Bar			Normal or Level of I	
☐	Finish Constraint	Remain Bar			Has Finish Constrai	△
☐	Dates on Bar	Current Bar			All Activities	
☑	Milestone	Current Bar			Milestone	◆ ◆
☑	% Complete	% Complete Bar			Normal	▬▬▬
☑	Summary	Current Bar			Summary	▬▬
☑	Float Bar	Float Bar			Not Started or In Pro	
☑	Neg Float Bar	Neg Float Bar			Negative Float	⊏▭▭

Bar Style	Bar Settings	Bar Labels

Grouping Band Settings

☐ Show bar when collapsed

☑ Show bar for grouping bands

8.3   Adding a Bar Label

If you wish to add a label, then it is best to create a bar that does not display the bar and only has a Label. This may be displayed against all activities and be easily displayed or hidden and saves editing multiple bars:

Display	Name	Timescale	User Start Date	User Finish Date	Filter	Preview
☐	Dates on Bar	Current Bar			All Activities	

Bar Style	Bar Settings	Bar Labels

Position	Label
Left	Start
Right	Finish

8.4 Baseline Bar Disappearing when a Activity is made an LOE

Often, after a Baseline has been set, there is a requirement to add more detail to a baselined activity but retain a view of the original Baseline bar. This may be achieved by converting the original Activity to an LOE and adding the detailed activities below the LOE and linking the new activities to the new LOE activity. If the original Baseline activity disappears, then you need to open the bars form and set the filter to **All Activities** and the Baseline bar will be displayed:

Bars					
Display	Name	Timescale	User Start Date	Filter	Preview
☐	Project Baseline Bar	Project Baseline Bar		All Activities	▭
☐	Project Baseline MS	Project Baseline Bar		Milestone ▽	▽

8.5 Why is there a gap in my Bar on the Data Date

When an activity is in progress the activity **Remaining Early Start** is set to the activity calendar Start Date & time,

❖ When the **Data Date** is set to Friday 17:00 there will be a gap between the **Data Date** line and the start of the **Remaining Bar**, Monday 08:00:

Finish	Remaining Early Start	Aug 25	Sep 01
		M T W T F S S	M T W T F S S
05-Sep-25 16:00	01-Sep-25 08:00		

❖ To prevent this set the **Data Date** to the Activity Calendar Start Date and Time, i.e. the **Remaining Early Start** date and time:

Finish	Remaining Early Start	Aug 25	Sep 01
		M T W T F S S	M T W T F S S
05-Sep-25 16:00	01-Sep-25 08:00		

NOTE: Microsoft Project normally defaults the **Status Date** to the end of a day, not the start.

8.6   No splits seen in an activity Actual bar with a Suspend and Resume

Oracle Activity Layouts do not neck and display a Suspend Date on an Actual Bar:

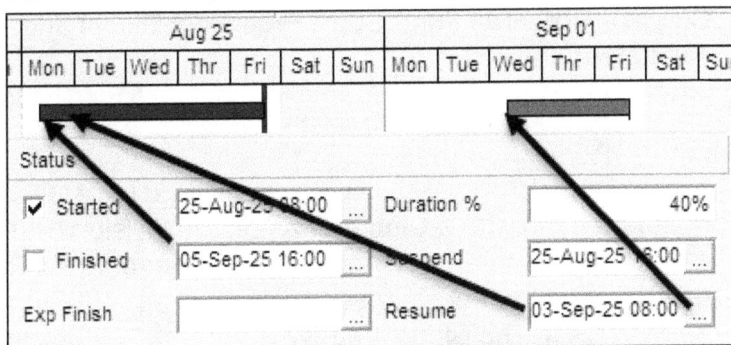

❖ To display the Suspend date on the bar you need to open the **Bars** form and select the **Actual Work** bar and check the **Bar Necking Settings, Activity nonwork intervals**:

❖ The result be as per below:

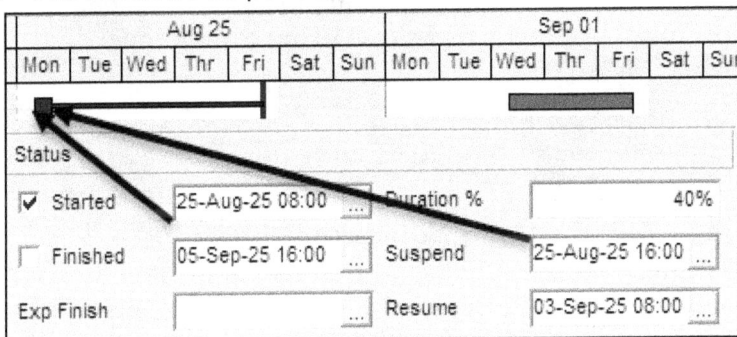

8.7   Displaying Free Float Bar

By default, P6 PPM will not display the **Free Float** bar and nor will **Visualizer**. If it is important to display it, there is an option using **UDF fields** and a **Global Change**:

❖ Create a **UDF Free Float Start**, **Free Float Finish** and **Free Float Value**:

Title	Data Type
Free Float Finish	Finish Date
Free Float Start	Start Date
Free Float Value	Number

❖ Populate them with a **Global Change**, the division by 8 changes hours to days. When different calendars are used, say 10 hour or several calendars, then you will need to revise the Global Change:

Then	Parameter	Is	Parameter/Value	Operator	Parameter/Value
	Free Float Start	=	Remaining Early Finish		
And	Free Float Value	=	Free Float	/	8.00
And	Free Float Finish	=	Remaining Early Finish	+	Free Float Value

❖ Then create a **User Bar** between the dates:

Display	Name	Timescale	User Start Date	User Finish Date	Filter
☑	Free Float	User Dates	Free Float Start	Free Float Finish	Not Started and In Progress

NOTE: It is important to set the filter as in the picture above, so a Total Float Bar is not displayed on completed activities.

8.8   Displaying All Baseline Bars

The Oracle Primavera **Bars** form **Default** option does not show either LOE or Summary bars, to show these bars you must format you bars as follows in the **Bars** form:

Display	Name	Timescale	Filter	Preview
☑	Primary Baseline	Primary Baseline Bar	All Activities	═══
☐	Primary Baseline	Primary Baseline Bar	Milestone	▽ ▽
☐	Primary Baseline	Primary Baseline Bar	Summary	═══

8.9   Displaying a Baseline Critical Path Bar

P6 by default does not display a **Baseline Critical** bar, but once you have set a Baseline you will find there are many more fields available in the columns form, including the **BL Total Float**, which may be used to draw a **Baseline Critical** bar.

To draw a **Baseline Critical** bar:

❖ Create but do not apply a Baseline Critical Filter, this will be used later:

Display all rows	Parameter	Is	Value	High Value
⊟	(All of the following)			
Where	BL Project Total Float	equals	0d	

❖ Create a Baseline Critical bar as per the picture below. It should be in the same position, which is set in the **Bar Style** tab, but below it in the Bars form, so it will be drawn over the top:

Display	Name	Timescale	User S	User F	Filter	Preview
☑	Project Baseline Bar	Project Baseline Bar			All Activities	▭
☑	Baseline Critical	Project Baseline Bar			Baseline Critical	▭

❖ The Baseline Critical bar will be drawn over the Baseline bar:

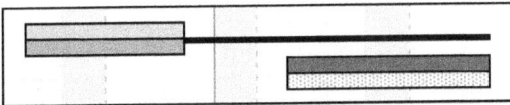

❖ **NOTE:** If you make the filter and layout Global, then any user may use the layout to display the Baseline Critical bar.

8.10 Formatting Columns

Column formatting is simple and may be completed by:

❖ Dragging columns to a new position and adjusting the width by dragging, or

❖ Opening the **Columns** form which is self-explanatory.

❖ It is often better to edit the column name so you may create narrower columns and still read the column name. For example, I always rename **Original Duration** to **Orig Dur**,

❖ **NOTE:** You will find that some columns, like the **Calendar** column may not be left formatted, only center or right.

8.11 Row Height and Show Icon

The height of all rows may be formatted by selecting:

❖ **View, Table Font and Row** to open the **Table, Font and Row** form.

❖ The options in this form are self-explanatory.

The height of a single row may be manually adjusted in a similar way to adjusting row heights in Excel, just grab the lower line of a cell and drag.

8.12 Moving and Rescaling the Timescale with the Mouse

Most formatting of the timescale may be completed with the mouse:

❖ To display hidden parts of the schedule the timescale may be grabbed and moved by placing the cursor in the top half of the Timescale. The cursor will turn into a 🖑; left click and drag left or right.

❖ The timescale may be rescaled, therefore increasing or decreasing the length of the bars and displaying more or less of the schedule, by placing the cursor in the bottom half of the Timescale. The cursor will turn into a 🔍 ; click, hold and drag left to make the bars shorter and right to make the bars longer.

❖ When there are no bars in view when you are viewing a time ahead or behind the activity dates, you may double-click in the **Gantt Chart** area to bring them back into view.

8.13   Inserting Attachments – Text Boxes and Curtain

Text Boxes and **Curtains** may be added to the Gantt Chart by right clicking in the Gantt Chart and selecting the required command.

❖ **A Text Box** will be attached and move with the activity.

❖ **Curtains** are useful for highlighting holidays etc. and are only displayed in the Layout they are created in:

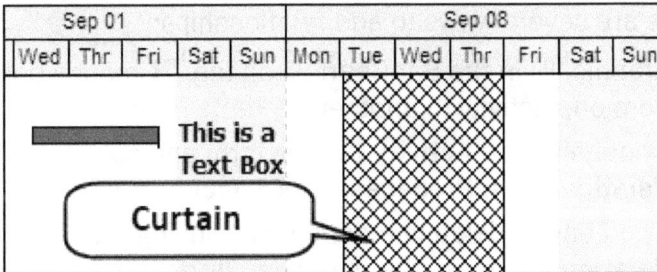

8.14   Line Numbers

Version 8.2 introduced a Microsoft Project style **Line Numbers**.

❖ Select **View, Line** Number to display or hide the Line Number.

❖ This is a very useful feature for reviewing a schedule to ensure that everyone in a meeting is looking at the same activity.

❖ But, as in Microsoft Project, this is an order and the number will change if the schedule is reordered.

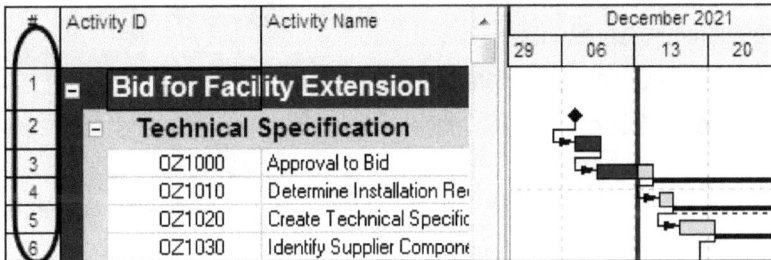

9 ADDING RELATIONSHIPS

Unlike Microsoft Project, P6 may have up to four relationships between two activities which makes creating a closed network simpler.

Microsoft Project only allows one relationship between two activities that makes creating a **Closed Network** more difficult, especially when utilizing **Ladder Scheduling**.

9.1 Adding relationships

There are several ways to add relationships:

❖ Graphically in the Bar Chart. Drag the ⌐↴ mouse pointer from one activity to another to create a dependency.

❖ By opening the **Activity Details** form and select the **Relationships, Predecessor** or **Successor** tabs.

❖ By editing or deleting a dependency using the **Edit Relationship** form, this is opened by double clicking on the relationship line.

❖ Opening the **Assign Predecessor** form or the **Assign Successor** form from the menu.

❖ By displaying the **Predecessor** and/or **Successor** columns.

❖ **Chain Linking** with a Finish-to-Start relationship. Select the activities in the order they are to be linked using the **Ctrl** key, right-click and select **Link Activities**.

9.2 Relationship Colors

The relationships may not be formatted but the color of the relationship represents:

❖ Red - Critical and therefore a Driving relationship,

❖ **Solid Black** - Non-Critical Driving relationship and therefore has Total Float,

❖ **Dotted Black** - Non-Critical Non-Driving relationship and has Free Float, and

❖ Blue - a selected relationship that may be deleted.

9.3 Hiding relationships

The relationships may be displayed or hidden by clicking the **Activity** toolbar ▣ icon or by checking and un-checking the **Show Relationships** box in the **Bar Chart Options** form, **General** tab.

9.4 Circular Relationships

A **Circular Relationship** is created when a loop is created in the logic. When you reschedule you will be presented with the **Circular Relationships** form, which identifies the loop.

NOTE: In some circumstances when **Retained Logic** is not used and there is progress assigned to one or more activities in the loop, then the **Circular Relationship** will not be identified.

NOTE: There are some interesting circumstances when a **Circular Relationship** is not recognized and occurs when:

❖ One activity in the loop is complete, and

❖ Either the **Tools**, **Schedule**, **Options** of **Progress Override** or **Actual Dates** are used:

NOTE: the schedule below is illogical, but P6 calculates:

NOTE: To avoid all the problems associated **Progress Override** or **Actual Dates** outlined in my other books I recommend only using **Retained Logic**.

9.5 Removing Relationships from Selected Activities

Should you wish to remove **ALL** the relationships between a group of activities the simplest way is:

❖ Select the activities in the **Activities** Window,

❖ Open the **Predecessor** form and select all activities in the form using the **Ctrl+A**,

❖ Then the **Delete icon** 🗑 will become available and you may delete all the **Predecessors** from the selected activities,

❖ Then open the **Successor** form and repeat to remove all the **Successors** from the selected activities,

❖ Now all the Predecessors and Successors will be deleted from the selected activities:

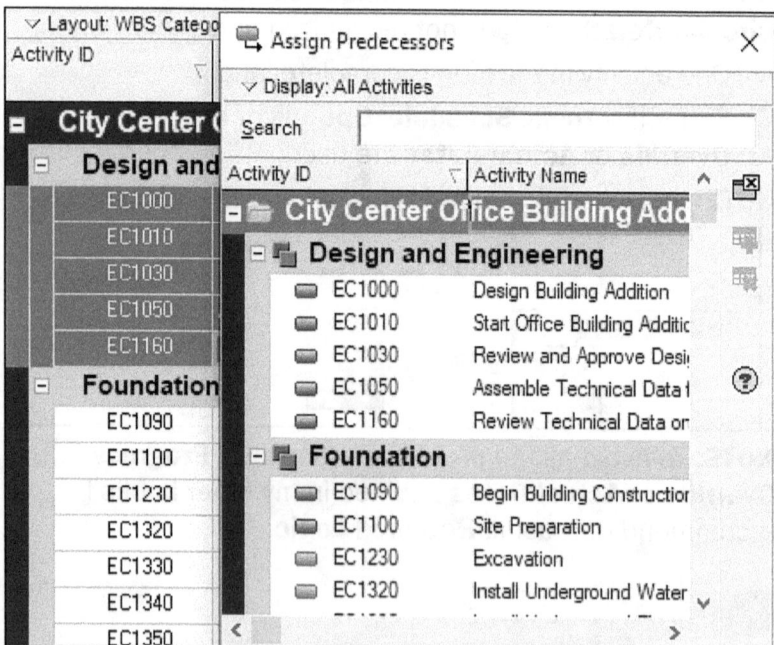

∨ Layout: WBS Catego	🖳 Assign Predecessors	✕
Activity ID	∨ Display: All Activities	
	Search	
■ **City Center**	Activity ID ▽	Activity Name ∧
⊟ **Design and**	■▤ **City Center Office Building Add**	
EC1000	⊟ 🖳 **Design and Engineering**	
EC1010	⊖ EC1000 Design Building Addition	
EC1030	⊖ EC1010 Start Office Building Additic	
EC1050	⊖ EC1030 Review and Approve Desi	
EC1160	⊖ EC1050 Assemble Technical Data 1	⑦
⊟ **Foundation**	⊖ EC1160 Review Technical Data or	
EC1090	⊟ 🖳 **Foundation**	
EC1100	⊖ EC1090 Begin Building Constructior	
EC1230	⊖ EC1100 Site Preparation	
EC1320	⊖ EC1230 Excavation	
EC1330	⊖ EC1320 Install Underground Water ∨	
EC1340		
EC1350	‹ ›	

9.6 Dissolving Activities and Retain Lag

When an activity is deleted then a chain of logical activities is broken. The **Edit**, **Dissolve** command and the right-click **Dissolve** command will delete an activity and joins the predecessors and successors with a Finish-to-Start relationship.

P6 Version 19 introduced the ability to retain lag when dissolving activities. When **Retain Lag** is selected from the **User Preferences**, **Calculations** tab, the dissolved activities predecessor and successor lags will be added together to calculate the lag in the new relationship:

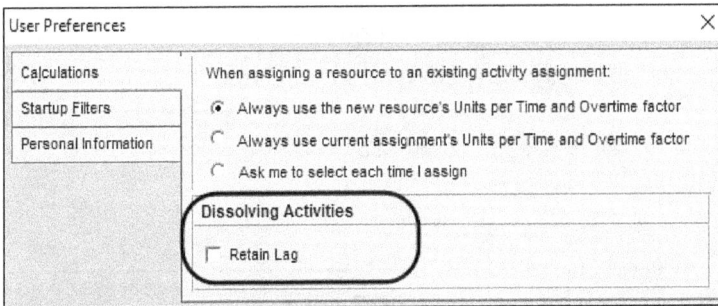

User Preferences	☓
Calculations	When assigning a resource to an existing activity assignment:
Startup Filters	⦿ Always use the new resource's Units per Time and Overtime factor
Personal Information	○ Always use current assignment's Units per Time and Overtime factor
	○ Ask me to select each time I assign
	Dissolving Activities
	☐ Retain Lag

- Before Dissolving Activity 2:

Activity Name	Orig Dur	Predecessor Details	Mar 31	Apr 07	Apr 14	Apr 21	Apr 28	May 05
Activity 1	5d							
Activity 2	5d	A1000: FS 5d						
Activity 3	5d	A1010: FS 5d						

- After Dissolving Activity 2 in P6 Version 19 without **Retain Lag** checked or in earlier versions of P6:

Activity Name	Orig Dur	Predecessor Details	Mar 31	Apr 07	Apr 14	Apr 21	Apr 28	May 05
Activity 1	5d							
Activity 3	5d	A1000: FS						

- After Dissolving Activity 2 in P6 Version 19 with **Retain Lag** checked:

Activity Name	Orig Dur	Predecessor Details	Mar 31	Apr 07	Apr 14	Apr 21	Apr 28	May 05
Activity 1	5d							
Activity 3	5d	A1000: FS 10d						

NOTE: The predecessor and successor lag are added together.

9.7   Reviewing Relationships, Leads and lags

In P6 Version 15 and earlier it is it is not possible to view all the leads and lags in columns from the user interface, as in Microsoft Project or Elecosoft (Asta) Powerproject.

Version 16.1 has added two more columns to the Activity View titled **Predecessor Details** and **Successor Details** allowing the checking of Lead and Lags in columns.

Relationships may be viewed in earlier versions of P6 using the following methods:

❖ Running the "Schedule Report - Predecessors & Successors", or

❖ As an export to Excel and this is a place where the leads and lags may be displayed in columns:

	A	B	C	L	M
1	pred_task_id	task_id	pred_type	lag_hr_cnt	delete_record_flag
2	Predecessor	Successor	Relationship Type	Lag(d)	Delete This Row
3	SH2002	SH2010	FS	-5	
4	SH2010	SH2020	FS	-5	
5	SH2020	SH2030	FS	-5	

9.8   Ladder Scheduling

Large negative lags are often considered unacceptable and Ladder Scheduling is a technique used to link a set of activities that have substantial overlap, such as pipe laying operations.

Most products allow multiple relationships between two activities, as per the P6 example below, where the activities are linked using two relationships, a SS+3d and a FF+3d:

Activity ID	Activity Name	Original Duration	Apr 01	Apr 08	Apr 15	Apr 22	Apr 29
A1000	Trench	15d					
A1010	Padding	15d					
A1020	Laying	15d					
A1030	Backfill	15d					

9.9   Relationship Comments

A **Comments** column may be added to the **Activities** window **Relationships, Predecessor** and **Successor** tabs in Version 20:

The **Comment** is seen in both the **Predecessor** and **Successor** window of each relationship assigned a comment:

❖ This is a very useful function and in the past one had to either use a Note or a UDF assigned to either the predecessor or successor activity to record notes about a relationship.

❖ This function is also useful for recording changes to relationships.

❖ The **Comments** column is available in the Activity Relationship report:

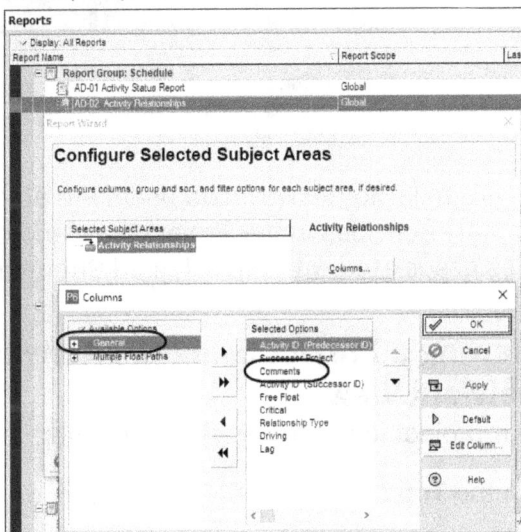

10 ACTIVITY NETWORK VIEW

The Activity Network, also known as the **PERT** View (Program (or Project) Evaluation and Review Technique), displays activities as boxes connected by the relationship lines.

Activities may be added, deleted and linked in this view in the same way as in the Gantt Chart View.

10.1 Viewing Relationships in Selected WBS Nodes

One of the main advantages of this view over the Gantt Chart is the ability to select two non-adjacent WBS Nodes and P6 will only display the activities associated with the selected nodes making it simple to check relationships in non-adjacent parts of the project.

NOTE: This is similar to the way Asta users are able to select Summary activities in the Project View and only see the activities associated with the selected Summary activities.

10.2 Viewing a Project Using the Activity Network View

To view your project in the **Network View** either:

❖ Click on the **Top Layout** toolbar 🔲 button, or

❖ Select **View, Show on Top, Activity Network**.

10.3 Formatting the Activity Boxes

Activity Boxes may be formatted from the **Activity Network Options** form, which may be displayed when the Activity Network View is displayed.

The formatting affects both the **Trace Logic** and **Activity Network Window** formatting for the layout that is being formatted:

❖ Select **View, Activity Network, Activity Network Options...**,or

❖ Right-click in the PERT area and select **Activity Network Options...**:

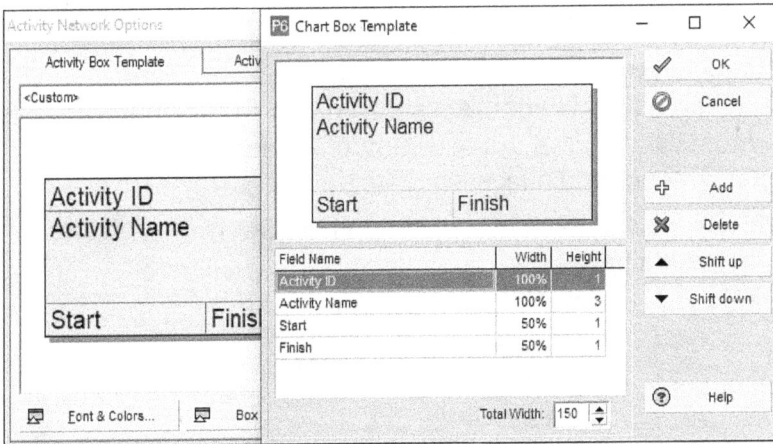

10.4 Reorganizing the Activity Network

Activities in the **Activity Network** view may be repositioned by dragging.

There are two functions available when right-clicking in the **Activity Network** view:

❖ **Reorganize** will reposition activities that have not been manually positioned, and

❖ **Reorganize All** will reposition all activities including those that have been manually positioned.

10.5 Saving and Opening Activity Network Positions

When activities are manually dragged into new positions on the screen for presentation purposes, it is possible to save and reload these positions after exiting the View.

These revised positions will be forgotten when you exit the **Activity Network View** and you must save the positions if you wish to see the activities in the same position again:

❖ **View, Activity Network, Save Network Positions...** will create an ***.ANP** file, and

❖ **View, Activity Network, Open Network Positions...** will enable an *.ANP file to be located and loaded which will reposition the activities as they were saved.

10.6 View Trace Logic

The Trace Logic option is displayed only in the bottom window and allows the selection of a number of predecessor and successor levels.

❖ This is achieved by selecting **View, Show on Bottom, Trace Logic**.

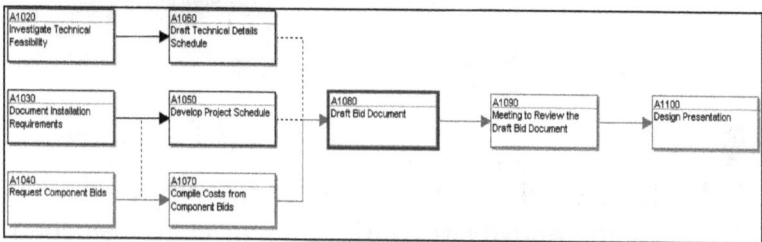

❖ To select the number of levels of predecessors to be displayed you are required to open the **Trace Logic Options** form.

❖ The form is then opened by right-clicking in the lower pane and selecting **Trace Logic Options.....**

11 CONSTRAINTS

Constraints are used to impose logic on activities that may not be realistically scheduled with logic links.

11.1 Assigning Constraints

When setting a constraint, the constraint time selected by P6 will often not be set at the start or finish of the activity calendar but set at 00:00 or some other irrelevant time.

Therefore, when setting constraints, you should **ALWAYS** display the time by selecting **Edit**, **User Preferences ...**, **Dates** tab to ensure the constraint time is compatible with the activity calendar.

Constraints may be assigned by:

❖ Displaying the appropriate columns,

❖ The **Activity Details** form,

❖ Dragging an Activity with the mouse,

❖ Typing a date into a **Start** or **Finish** column.

11.2 Number of Constraints per Activity

Two constraints are permitted against each activity.

They are titled **Primary** and **Secondary Constraint**.

After the Primary has been set, a Secondary may be set only when the combination is logical and therefore, a reduced list of constraints is available from the Secondary Constraint list after the Primary has been set.

11.3 As Late As Possible Constraint

In P6 this is not an **As Late As Possible Constraint** but a **Zero Free Float** constraint. This will only delay an activity within its Free Float and not delay any successor activity. A true **As Late As Possible Constraint** would delay all successor activities with **Total Float**, as in Microsoft Project.

Importing a project from Microsoft Project with this constraint may calculate very differently in P6.

11.4 Expected Finish Constraint

This constraint will only work if the **Tools**, **Schedule...**, ▷ Options... , **Use Expected Finish dates** check box is checked.

This constraint is set in the **Status** section of the **Activity Details**, **Status** tab, not under the **Constraints** section as one would expect.

An **Expected Finish** constraint recalculates the **Remaining Duration** on a permanent basis.

The author uses an **Expected Finish** constraint as a tool to calculate the **Remaining Duration** when the planned finish date is known, but then removes it after calculation. This is because when the **Expected Finish** constraint date is in the past it will zero out the **Remaining Duration**.

11.5 Constraint Identification

After a constraint is set the date will have an asterisk "*" next to it.

❖ Start Constraints will have the "*" next to the Start Date, and

❖ Finish Constraints will have the "*" next to the Finish Date.

11.6 External Dates

Activities with **External Early Start** and **External Late Finish** dates will display an "*" by the date, but the constraint will not be displayed in **Details**, **Status** tab.

External Dates are constraints created when a project is exported from Primavera Contractor and/or another P6 database and imported into P6. They act like Early Start and Late Finish Constraints and are used to represent the relationships that would have originally provided the Early Start and Late Finish dates to the Critical Path calculations of the imported schedule.

You should ALWAYS check for **External Dates** after importing a project

11.7   Mandatory Constraints

Mandatory constraints must be avoided as they override network logic and then float may not be correctly calculated on activities assigned a **Mandatory Start** or **Mandatory Finish** constraint. At that point in time you will no longer have a **Critical Path** schedule as required by many contracts.

The **Bid Documents Submitted** has a **Mandatory Finish** constraint assigned 4 days before its predecessor:

Bid Document		-4d	
OZ1100	Create Draft of Bid Document	7d	
OZ1110	Review Bid Document	-4d	
OZ1120	Finalise and Submit Bid Document	-4d	
OZ1130	Bid Document Submitted	0d	

An activity with a **Mandatory** constraint date earlier than a predecessor will ignore all predecessors and be scheduled earlier than logic will normally allow and will display zero **Total Float**. This situation, in the picture above, is illogical.

When you have dates that an activity must take place on, then it is better to use a **Must Start** or **Must Finish** constraint as **Negative Float** will be generated on the activity in question and the issue will be clearly seen.

11.8   Activity Notebook

It is often important to note why constraints have been set. Primavera has functions that enable you to note information associated with an activity, including the reasons for setting the constraint.

The **Activity Details** form has a **Notebook** tab, which enables notes to be assigned to **Notebook Topics** and has some word processing-type formatting functions.

Notebook Topics are created by selecting **Admin, Admin Categories...** when opening a PPM database and through the Web Tool Administration area when opening a EPPM database.

12 GROUP, SORT AND LAYOUTS

Group and Sort enables data such as:

❖ Activities in the **Activities Window**,

❖ WBS Nodes in the **WBS Window**,

❖ Projects in the **Project Window**,

to be sorted and organized under other parameters, such as **Dates** and **Resources** or user defined **Activity** and **Project Codes**.

❖ **Layouts** is a function where the formatting of parameters such as the **Group and Sort**, **Columns** and **Bars** are saved and reapplied later.

❖ A **Layout** may be edited, saved, or reapplied at a later date and may have a **Filter** associated with it.

❖ Layouts may be exported and imported into another database.

❖ Layouts contain the formatting for all options of both the top and bottom pane, including columns displayed in the **Details** lower pane window, **Activity Usage Spreadsheet**, **Activity Usage Profile**, **Resource Usage Spreadsheet**, **Resource Usage Profile** settings

❖ Layouts do not include the **Date** and **Time** formatting and are set the **User Preferences**.

12.1 Activity Groups and Sort form

To Group and Sort activities, open the **Group and Sort** form by:

❖ Clicking the [icon] toolbar icon, or

❖ Select **View, Group and Sort by**, **Customize**.

Here is a quick summary of the functions:

❖ **Show Group Totals**, when unchecked hides the summary data in the bands, which prevents the truncating of Band titles.

Summary Data Displayed

Activity ID	Original Duration	Start	Finish	Total Float
■ **Bid for Fac**	139d	08-Dec-09 A	22-Jan-10	0d
⊟ **Research**	31d	08-Dec-09 A	21-Dec-09	6d
OZ1000	0d	08-Dec-09 A		
OZ1010	1d	08-Dec-09 A	08-Dec-09 A	
OZ1020	8d	09-Dec-09 A	21-Dec-09	2d
⊟ **Estimate**	52d	22-Dec-09	08-Jan-10	3d
OZ1070	2d	07-Jan-10	08-Jan-10	0d

Summary Data Hidden

Activity ID	Original Duration	Start	Finish	Total Float
■ **Bid for Faciliity Extension**				
⊟ **Research**				
OZ1000	0d	08-Dec-09 A		
OZ1010	1d	08-Dec-09 A	08-Dec-09 A	
OZ1020	8d	09-Dec-09 A	21-Dec-09	2d
⊟ **Estimate**				
OZ1070	2d	07-Jan-10	08-Jan-10	0d

❖ **Show Grand Totals** provides a total of all the activities in a band at the top of a view. This is useful when multiple projects are open, and you wish to see a total of all projects.

Activity ID	Activity Name	BL Project Total Cost	BL Project Labor Units
■ **Total**		A$6,899,980.56	100397
⊞ **Nesbid Building Expansion**		A$550,470.40	9346
⊞ **Haitang Corporate Park**		A$636,980.80	10735
⊞ **City Center Office Building Addition**		A$1,162,028.80	20110
⊟ **Harbour Pointe Assisted Living Center**		A$4,550,500.56	60201

❖ **Show Summaries Only** hides all the activities and displays only the WBS or Codes that have been used to summarize the activities.

Activity ID	Activity Name	Early Start	Early Finish	BL Budgeted Labor Cost	, 2003	Qtr 1, 2004	Qtr 2, 2004
					ov Dec	Jan Feb Mar	Apr May Jun
■ **1-5 OzBuild Bid**		01-Dec-03	29-Jan-04	$22,830	▼━━━━━━▼ 29-Jan-04, 1-5 OzBuild Bid		
1-5.1 Research		01-Dec-03	17-Dec-03	$6,640	▼━▼ 17-Dec-03,1-5.1 Research		
1-5.2 Estimation		12-Dec-03	15-Jan-04	$8,150	▼━━▼ 15-Jan-04, 1-5.2 Estimation		
1-5.3 Proposal		15-Jan-04	29-Jan-04	$8,040	▼━▼ 29-Jan-04, 1-5.3 Proposal		

❖ **Shrink vertical grouping bands** narrows the Vertical Bands on the left of the screen. This is useful in projects with a number of levels in the WBS as this provides more usable screen space and paper width for printing.

Option Unchecked   **Option Checked**

Activity ID	Activity Name	Original Duration
Bid for Facility Extension		
Research		
0Z1000	Bid Request Docume...	0.0d
0Z1010	Bid Strategy Developed	1.0d
0Z1020	Technical Feasibility S...	8.0d
Estimate		

Activity ID	Activity Name	Original Duration
Bid for Facility Extension		
Research		
0Z1000	Bid Request Docume...	0.0d
0Z1010	Bid Strategy Developed	1.0d
0Z1020	Technical Feasibility S...	8.0d
Estimate		

❖ The **Group By** box is used to create bands. The **Group By** column selects the data to be used to group activities. Unchecking **Indent** prevents a hierarchical view. **To Level** allows the truncation of hierarchical fields like WBS so additional data fields may be added below. Group by is used for fields like Start Date. The pictures below give a snapshot of the way this form may be used:

Group By	Indent	To Level	Group Interval	Font & Color
Early Start			Week	**12 Arial**
WBS	☐	All		**11 Arial**
Resources				9 Arial
				8 Arial

Activity ID	Activity Name	Early Start	Early Finish
01-Dec-03		01-Dec-03	11-Dec-03
Research		01-Dec-03	11-Dec-03
Sales Engineer, System Engin...		01-Dec-03	01-Dec-03
A1010	Bid Strategy Meeting	01-Dec-03	01-Dec-03
System Engineer		02-Dec-03	11-Dec-03
A1020	Investigate Techni...	02-Dec-03	11-Dec-03
08-Dec-03		12-Dec-03	15-Jan-04
Research		12-Dec-03	17-Dec-03
Sales Engineer, System Engin...		12-Dec-03	17-Dec-03
A1030	Document Installati...	12-Dec-03	17-Dec-03

© *Eastwood Harris*

❖ **Sort Banding Alphabetically** means sort the WBS by the code value and is useful for giving the WBS two sort orders:

WBS Code	WBS Name
⊟ 📷 OZB	Bid for Facility Extension
⬛ OZB.C	Technical Specification
⬛ OZB.A	Delivery Plan
⬛ OZB.B	Bid Document

When unchecked they are sorted naturally per the picture above:

Activity ID	Activity Name
⊟ **OZB-04 Bid for Facility Extension**	
OZB-04.C	Technical Specification
OZB-04.A	Delivery Plan
OZB-04.B	Bid Document

When this is checked the bands are sorted by the code assigned to the Activity Code or WBS Code:

Activity ID	Activity Name
⊟ **OZB-04 Bid for Facility Extension**	
OZB-04.A	Delivery Plan
OZB-04.B	Bid Document
OZB-04.C	Technical Specification

❖ **Hide if empty** - Check this box to hide bands that:

➢ Have not been assigned an activity, or

➢ When activities have been filtered out and only the bands remain.

❖ **Show Title, Show ID/Code and Show Name/Description** These options format the display of the band title. It is not possible to uncheck all the options as there would then not be a title in the band.

12.2 Sorting

The [🔲 Sort...] icon opens the **Sort** form where the order of the activities in each band may be specified.

NOTE: This order may be easily overridden by clicking on the column titles to reorder activities and therefore the use of this option is problematic, as clicking on the column header is very simple and will override options set here.

12.3 Group and Sort Projects at Enterprise Level

Projects in the **Projects Window** may be Grouped and Sorted in a similar way to activities.

When a database is opened, the projects are by default displayed under the Enterprise Project Structure (EPS) in the **Projects Window**.

The projects may be Grouped and Sorted under a number of different headings by using the **Group and Sort** form in conjunction with **Project Codes** or **Project UDFs**.

12.4 Understanding Layouts

Layouts is a function in which the formatting of parameters such as the Group and Sort, Columns and Bars is saved and reapplied later.

A **Layout** may be edited, saved, reapplied later and may have a **Filter** associated with it. Layouts contain the formatting for all options of both the top and bottom pane.

Layouts do not save the Date, Time and Duration formatting, these are set in the **User Preferences** form.

Although **Group and Sort** is available in many forms, Layouts are only available in a few places including the following Windows:

❖ Projects

❖ WBS

❖ Activities

❖ Tracking

12.5 Copying Layouts from one database to another

The **Open Layout** form allow Layouts to be exchanged between P6 databases using the **Import** and **Export** commands. The XML file import and export also will allow **Activity Layouts** to exchanged between database. The Layout formatting is saved in a Layout file which has a ***.PLF** extension.

12.6 Removing/Reordering Bottom Pane tabs

Often a Layout will have many tabs displayed that you will not use or are in an illogical order.

It is recommended that you remove all the tabs that you are not using so that the screen is not so busy, and it is simpler to find commands

Right-click any tab in the top of the details pane and select **Customize Project Details....** to remove or reorder bottom pane tabs. The arrows are used to hide and display tabs and reorder them.

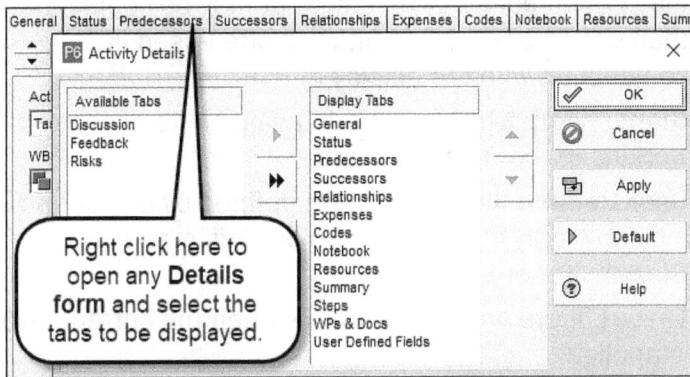

General	Status	Predecessors	Successors	Relationships	Expenses	Codes	Notebook	Resources	Summ

P6 Activity Details ✕

Available Tabs
Discussion
Feedback
Risks

Display Tabs
General
Status
Predecessors
Successors
Relationships
Expenses
Codes
Notebook
Resources
Summary
Steps
WPs & Docs
User Defined Fields

✓ OK
⊘ Cancel
Apply
▷ Default
⑦ Help

Right click here to open any **Details** form and select the tabs to be displayed.

12.7 Disable Auto-Reorganization Shft+F12

When you reassign a WBS or Activity Code value then the activities may reorganize and you will end up with your cursor in a different part of the project.

Often it is undesirable for the activities to reorganize until you have completed the recoding of a number of activities and in this situation, it is best to **Disable Auto-Reorganization**, found under the **Tools** menu.

13 FILTERS

13.1 Understanding Filters

Primavera has an ability to display activities that meet specific criteria. You may want to see only the incomplete activities, or the work scheduled for the next couple of months or weeks, or the activities that are in progress.

Primavera defaults to displaying all activities. There are a number of pre-defined filters available that you may use or edit. You may also create one or more of your own.

A filter may be applied to display or to highlight only those activities that meet a criterion. Unfortunately, the **Highlight** function does allow the normal functions of printing the highlighted activities, reversing highlighting or hiding highlighted activities.

13.2 Filter Types

There are four types of filters::

❖ **Default** filters which are supplied with the system and may not be edited or deleted, but may be copied and then edited or modified and are often used in conjunction with the display of bars.

❖ **Global** filters which are made available to anyone working in the database, and

❖ **User Defined** filters which are defined by a user and available only to that user, unless it is made into a **Global** filter or **Layout** filter,

❖ **Layout** filters are only available when the current layout is applied.
NOTE: If the current layout is a **Project** layout then this effectively makes the **Layout** filter a project filter.

The following types of filters are not available:

❖ **Drop down** or **Auto filters** as in Excel and Microsoft Project.

❖ **Interactive** filters as available in SureTrak and Microsoft Project. This is when a filter is applied, and the user is offered choices from a drop-down list. The lack of this

© *Eastwood Harris*

function may result in an excessive quantity of filters being generated or the user continually editing frequently used filters.

❖ **Project Filters** which the lack off often results in a proliferation of Global filters so all users of a project can access the filters designed for a specific project. This creates an ongoing maintenance task for the database administrator, removing old filters after the project is complete.

NOTE: It is recommended that in large databases a coding system is set up for Global filters when users are allowed to create them or users use Layout filters .

13.3   Applying a Filter

Filters are applied from the **Filters** form which may be opened by:

❖ Clicking on the [▽] icon, or

❖ Selecting **View, Filters...**, Customize...or

❖ Right-clicking in the columns area and selecting **Filters...**

NOTE: If the **All Activities** check box in the **Filter** form is not checked then there is a filter applied.

13.4   Creating a New Filter

Filters may be created from the **Filters** form and are fairly self-explanatory.

To see how filters are constructed you may either:

❖ Copy some of the existing Global filters and look at them,

❖ Look at other filters if they exist, or

❖ Buy one of my other P6 books.

13.5   Understanding Resource Filters

When filtering on resources, the filter must use the option of **contains** in the **Is** column and not **equals**. When an activity has been assigned more than one resource, then the activity will not be selected with a filter using the **equals** parameter.

13.6 WBS Filters

Filters created for a WBS Code will only work on the project they are created in because the WBS Code includes the Project ID.

NOTE: If your project uses WBS filters, then you should not change the Project ID of your Current or Work In Progress project. When you copy your project, then the copies should be assigned a new Project ID and the project you are working on should always have the same Project ID.

13.7 Some Useful Ideas for Filters

Here are some things you can do with filters:

❖ Find SS and FF relationships by using the **Predecessor** and **Successor** contain **SS** or **FF**. This is useful when used in conjunction with highlighting.

Display all rows	Parameter	Is	Value	High Value
⊟	(Any of the following)			
Where	Predecessor Details	contains	SS	
Or	Predecessor Details	contains	SS	
Or	Successor Details	contains	FF	
Or	Successor Details	contains	FF	

(Display: Filter)

❖ Isolate activities with a specific resource by using the **Resource Name** plus **contains**. If you use **Resource Name** plus **equals,** you will not see any activity that has more than one resource assigned.

Display all rows	Parameter	Is	Value	High Value
⊟	(All of the following)			
Where	Resource IDs	contains	PM	

❖ When an activity name contains a word, the text that is entered is not text sensitive, so the filter below will isolate activities with Bid or bid in the name:

Display all rows	Parameter	Is	Value	High Value
⊟	(All of the following)			
Where	Activity Name	contains	bid	

❖ Creating a look-ahead filter, filtering on the following parameters:

Display all rows	Parameter	Is	Value	High Value
⊟	(Any of the following)			
⊟ Where	(All of the following)			
Wher	Activity Status	equals	In Progress	
⊟ Or	(All of the following)			
Wher	Start	is greater than	CD	
And	Start	is less than	DD+2W	

NOTE: The **Look-ahead** filter in the Oracle sample database does not work.

13.8 Exporting Filters

Filters may not be exported on their own, but if you wish to export a filter, then it may be made a **Layout** filter by using the **Copy As Layout** command in the **Filters** form and then have the Layout exported:

This filter has been made a Layout Filter

13.9   Activity Critical Path Visibility Toolbar

There is a new toolbar in Version 21 titled **Activity Critical Path**.

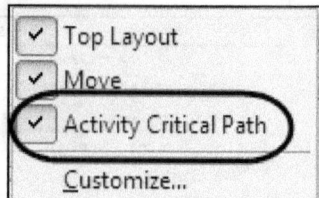

❖ The toolbar is displayed by right clicking on any toolbar and checking the **Activity Critical Path** option:

❖ This will display the toolbar:

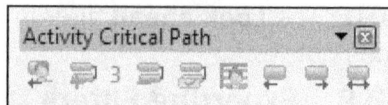

❖ IMPORTANT NOTE:
Clicking on a button will activate the command and clicking again will deactivate the command

We will work through the icons one at a time and explain how the functions work.

13.9.1   Backward Resource Driven Critical Path

The schedule below has no relationships between activities and has been levelled.

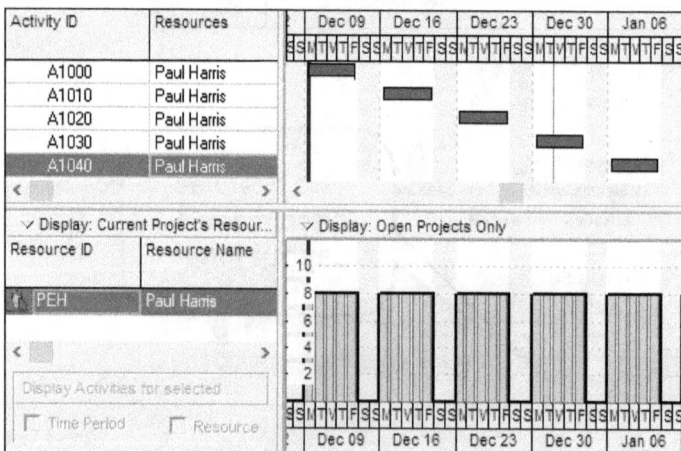

When selecting A1040 and **Backward Activity Critical Path** you will see the result below:

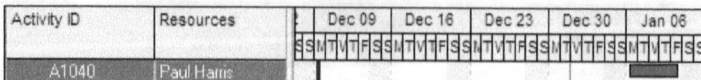

When selecting A1040 and **Backward Resource Driven Critical Path** you will see the result below:

Activity ID	Resources	Dec 09	Dec 16	Dec 23	Dec 30	Jan 06
A1000	Paul Harris					
A1010	Paul Harris					
A1020	Paul Harris					
A1030	Paul Harris					
A1040	Paul Harris					

NOTE: Review the **Backward Activity Critical Path** now and then come back and review this one.

13.9.2 Increase Maximum Multiple Critical Path Counts

This increases the value of the **Maximum Multiple Critical Path Counts** which is displayed in the button below.

13.9.3 Maximum Critical Path Count

This is not a command button but displays the number of **Maximum Critical Paths** that has been set by the user.

Selecting one **Maximum Critical Path Count:**

Activity ID	Activity Name	2011			2012			
		Q2	Q3	Q4	Q1	Q2	Q3	Q4
EC1290	Fabricate and Deliver Heat Pump and Controls							
EC1650	Set Heat Pump							
EC1660	Connect Equipment							

Selecting two **Maximum Critical Path Count:**

Activity ID	Activity Name	2011			2012			
		Q2	Q3	Q4	Q1	Q2	Q3	Q4
EC1290	Fabricate and Deliver Heat Pump and Controls							
EC1650	Set Heat Pump							
EC1660	Connect Equipment							
EC1350	Concrete Foundation Walls							
EC1360	Form and Pour Slab							
EC1370	Backfill and Compact Walls							
EC1390	Erect Structural Frame							
EC1420	Floor Decking							
EC1430	Concrete First Floor							
EC1540	Structure Complete							
EC1480	Concrete Second Floor							
EC1550	Brick Exterior Walls							
EC1620	Building Enclosed							
EC1600	Insulation and Built-up Roofing							
EC1640	Install Wiring and Cable							

Thus, this command determines how many chains of events you wish to see.

13.9.4 Decrease Maximum Multiple Critical Path Counts

This decreases the value of the **Maximum Multiple Critical Path Counts** which is displayed in the button above.

13.9.5 Select Maximum Multiple Critical Path Counts

This opens a form allowing the setting of the **Maximum Multiple Critical Path Counts**:

13.9.6 Enable Group and Sort

Clicking this button hides and displays the current applied Group and Sort bands. Hiding the bands enables the user to see the chain of events as an uninterrupted single chain of activities:

❖ Group and Sort enabled:

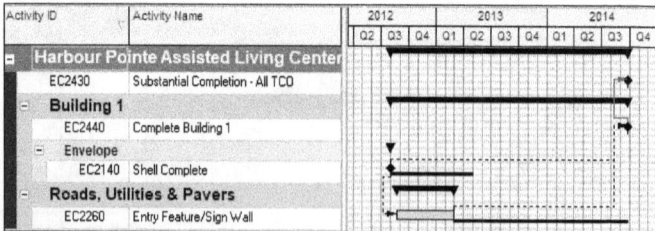

❖ Group and Sort disabled:

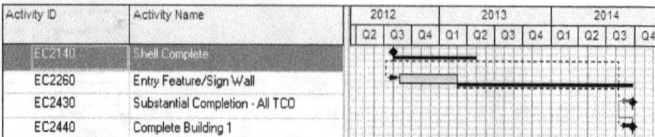

13.9.7   Backward Activity Critical Path

This option displays the chain of events before a selected activity that has the least amount of Float. If there is a critical path it will display this and if there is not, then it will find the chain of events with the lease amount of float to the start of the project and display this as the **Backward Activity Critical Path**.

The picture below displays the **Backward Activity Critical Path** from activity EC1500, which is a chain of events that is on the schedule Critical Path.

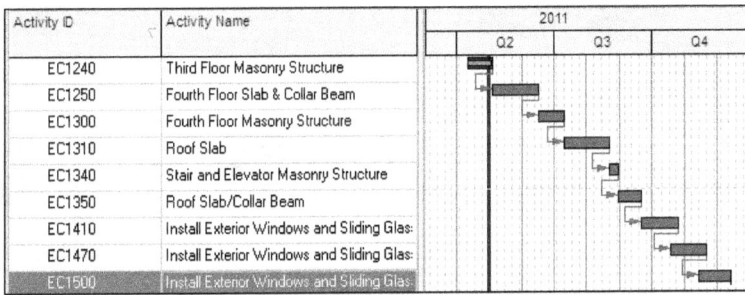

Activity ID	Activity Name	2011		
		Q2	Q3	Q4
EC1240	Third Floor Masonry Structure			
EC1250	Fourth Floor Slab & Collar Beam			
EC1300	Fourth Floor Masonry Structure			
EC1310	Roof Slab			
EC1340	Stair and Elevator Masonry Structure			
EC1350	Roof Slab/Collar Beam			
EC1410	Install Exterior Windows and Sliding Glass			
EC1470	Install Exterior Windows and Sliding Glass			
EC1500	Install Exterior Windows and Sliding Glass			

13.9.8   Forward Activity Critical Path

This option displays the chain of events after a selected activity that has the least amount of Float. If there is a critical path it will display this path and if there is not, then it will find the chain of events with the lease amount of float to the end of the project and display this as the **Forward Activity Critical Path**.

The picture below displays the **Forward Activity Critical Path** from activity EC1840, which is a chain of events that is not on the schedule Critical Path.

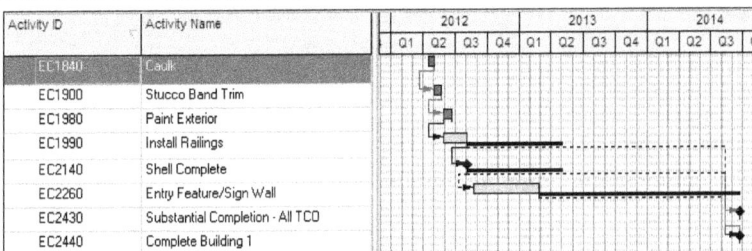

Activity ID	Activity Name	2012				2013				2014			
		Q1	Q2	Q3	Q4	Q1	Q2	Q3	Q4	Q1	Q2	Q3	Q
EC1840	Caulk												
EC1900	Stucco Band Trim												
EC1980	Paint Exterior												
EC1990	Install Railings												
EC2140	Shell Complete												
EC2260	Entry Feature/Sign Wall												
EC2430	Substantial Completion - All TCO												
EC2440	Complete Building 1												

13.9.9 Activity Critical Path

This option displays the chain of events before and after a selected activity that has the least amount of Float, as per the parameters above.

Thus, the picture below shows both the **Forward Activity Critical Path** and **Backward Activity Critical Path** from Activity PD1040.

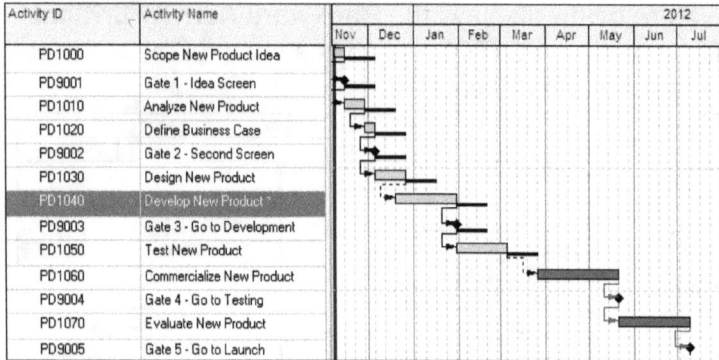

Activity ID	Activity Name										2012	
		Nov	Dec	Jan	Feb	Mar	Apr	May	Jun	Jul		
PD1000	Scope New Product Idea											
PD9001	Gate 1 - Idea Screen											
PD1010	Analyze New Product											
PD1020	Define Business Case											
PD9002	Gate 2 - Second Screen											
PD1030	Design New Product											
PD1040	Develop New Product *											
PD9003	Gate 3 - Go to Development											
PD1050	Test New Product											
PD1060	Commercialize New Product											
PD9004	Gate 4 - Go to Testing											
PD1070	Evaluate New Product											
PD9005	Gate 5 - Go to Launch											

© *Eastwood Harris*

14 PRINTING AND REPORTS

14.1 Output Options

There are several tools available to output your schedule:

❖ The **Printing** function prints the data displayed in the current Layout.

❖ The **Reporting** function prints reports, which are independent of the current Layout. Primavera supplies a number of predefined reports that may be tailored to suit your own requirements. Reports will not be covered in detail in this publication.

❖ The **Project Web Site Publisher** to publish the schedule to a web site using the **Tools**, **Publish** command. This is a very powerful way of getting a large amount of data into a website, simply and easily.

❖ You may use **Visualizer** which was new to P6 Version 8.3 and an update of the P6 Version 8.2 **Timescaled Logic Diagram** module. It is a separate piece of Windows Client software that reads your Oracle Primavera database and allows a higher level of Gantt Chart customization than available from the Activities Window, in either Professional or Web Client. Version 16 added **Claim Digger** to Visualizer and named it **Schedule Comparison**.

❖ You may also copy and paste text data from columns and some tables into Excel and other products.

❖ If you are working in an **EPPM** database, then the **Web client** has a large number of powerful reporting options outside the scope of this book

14.2 Management Considerations

❖ Each time you report to the client or management, it is recommended that you save a copy of your printout or report, and a pdf file is an excellent method of saving this data. In conjunction with a robust file-naming convention that includes the project title and Data Date, a pdf file will enable you to reproduce these reports at any point in time in the future and have available a copy of the project schedule for dispute resolution purposes.

❖ It is good practice to keep a copy of the project after each update, especially if litigation is a possibility. A project may be copied either by creating a Baseline, by exporting the project as an XML file (which will also save any Baseline) or XER (without a Baseline) or by using the project copy function, then making the Status inactive in the General tab of the Projects Window. It is important to note that although the project may be marked as inactive it may still be opened and modified.

❖ It is often better to keep Layouts just for reporting, that do not get messed up when working on a project and layouts for working that you may edit from time to time

❖ If your organization has limited funds then you may consider investing in a product like **Schedule Reader http://www.schedulereader.com** which allows people without access to P6 to reproduce reports form P6 export files in a similar format.

14.3   Printing

When a Layout is split, the lower pane may be printed with the upper pane, with the exception of the **Activity Details** pane that may not be printed.

Print settings, such as headers and footers, are applied to the individual Layouts and the settings are saved with that Layout.

NOTE: Layouts do not save the Date, Time and Duration formatting, these are set in the **User Preferences** form.

Printing with P6 is very simple and a lot better than some of the competitors and here are some notes to guide you through the process:

The **Page** and **Margins** tabs are self-explanatory, but the unit of measurement in the **Margins** tabs is inches.

Headers and **Footers** are formatted the same way.

❖ **Divide Into:** – determines the number of sections the Header/Footer is divided into from 1 to 5 sections.

❖ **Include on:** – determines on which pages the Header/Footer is to appear: First Page, Last Page, All Pages, or No Pages.

   NOTE: I usually put things I want to see on every page, like project name, page number and Data Date, in the header and show this on every page, and things I want to see once, like the logo and legend, in the footer and print on the first page only.

❖ **Height:** – enables the user to select the height of the Header/Footer.

❖ **Define Footer**

   ➢ **Show Section Divider Lines** check box – hides or displays the divider lines between the sections.

   ➢ The sections may be sized by manually moving the divider lines with the mouse and the slide underneath the **Show Section Divider Lines**.

❖ **Section Content**

➢ This may be selected by clicking on the [▾] icon under the **Section** title and a subject type to be displayed selected.

➢ **(None)** – leaves the section blank.

➢ **Gantt Chart Legend** – displays all the bars checked in the display column of the **Bars** form and only the fonts may be edited by clicking on the little [Font] icon at the bottom.

➢ **Picture** – enables a picture to be placed in the footer and it may be manually adjusted to fit the space or automatically adjusted by checking the **Resize picture to fit the selection** box.
NOTE: The picture directory must be available to the user for the picture to be displayed and the **Stored Images** function overcomes this issue because the pictures are stored in the database.

➢ **Revision Box** has a **Revision Box Title:** – the following information may be entered manually: Date, Revision, Checked, Approved.

➢ **Stored Images** – New to P6 Version 18. Up to 20 pictures at a max of 500x500 dpi may be saved in a database and are available for any user to access. Pictures are uploaded using **Enterprise, Store Image....**

➢ **Text/Logo** – enables many types of data to be displayed including text, a data item selected from the drop down box, fonts formatted by clicking on the formatting icons [A] [≡ ≡ ≡] [≔ ≔] [◂≣ ▸≣], a Logo inserted by clicking on the [🖼] icon, Tables added by clicking on the [▦] icon, and a Hyperlink added by clicking on the link icon [🖉] which opens the **Hyperlink** form.

➢ **NOTE:** When you are editing text in a header or footer and a line return created by using the **Enter**

Key will result in a paragraph break and a large gap between the lines. This may be prevented by using a soft return using the **Shift & Enter** command:

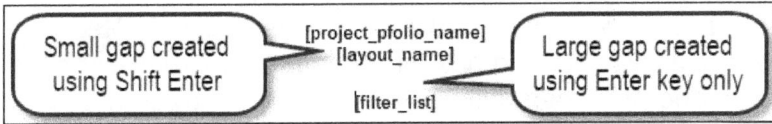

Small gap created using Shift Enter	[project_pfolio_name] [layout_name] [filter_list]	Large gap created using Enter key only

The **Options** tab:

❖ **Timescale Start** and **Timescale Finish** are useful to ensure the required area of the Gantt Chart is printed and to ensure text at the end of a bar is not truncated.

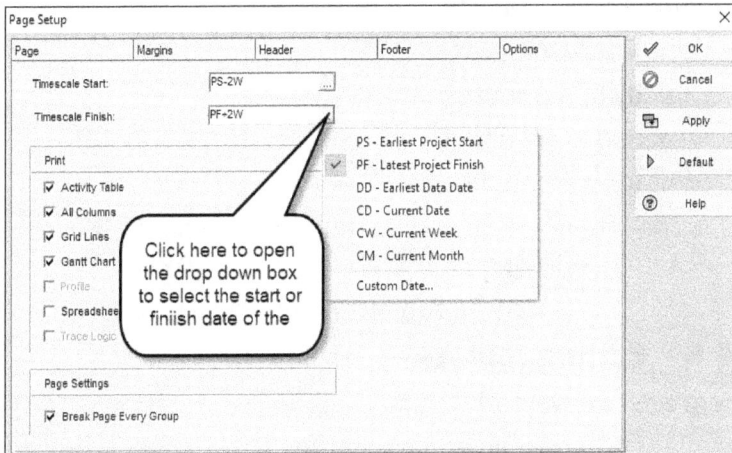

❖ A Histogram may be selected to be printed, and

❖ Automatic Page Breaks may be set here as well as in the **Group and Sort** form.

➢ The option in this form only works when there is more than one level 1 band in the **Group and Sort** form, and would not work for a WBS when Indented to All Levels, but would work when activities are grouped by a Start or Finish date.

➢ The **Group and Sort** form allows more control over where page breaks are inserted using the **Page Break** option.

14.4 Not able to type a "S" in the Header or Footer

When adding text in some versions of P6 Version 8, the letter "s" may not be typed into the Header and Footer. You will be able to cut and paste the required text with an "s" from another program like Notepad or you may create an image and insert the image into the header or footer. This issue was rectified in P6 Version 15.

14.5 Running Reports

There are a large number of reports that may be run by selecting **Tools**, **Reports** and output to:

❖ Print Preview,

❖ Directly to a printer,

❖ Attached to an email,

❖ Sent to an HTML file, or

❖ Made into an ASCII text document and usually they Automatically open in Excel.

Multiple reports may be run by creating a **Batch**.

14.6 Editing reports

There are two ways of editing reports:

❖ All reports may be edited with the **Report Editor**, which is hard work to learn; right click on the report and select **Modify**.

❖ Some reports, which have a ☐ icon (that looks like an ice cream cone) may be edited using the **Report Wizard**. Select the report and then select **Tools**, **Report Wizard**.
   NOTE: The **Report Wizard** is a lot simpler to use than the **Report Editor**.

14.7 Publish to a Web Site

Primavera has several functions that enable a project to be published to a web site, which is effectively the only "Free Reader" that Oracle Primavera provides with P6.

The **Tools**, **Publish**, menu has three options for creating a web site for a currently opened project:

❖ **Project Web Site...** creates a complete web site with any Reports or Layouts that have been created. This is a very useful function if it is required to publish a large amount of data.

❖ **Activity Layouts...** creates a web site with just the selected Activity Layout.

❖ **Tracking Layouts...** creates a web site with just the selected Tracking Layout.

14.8 Understanding Visualizer

Visualizer was new to P6 Version 8.3 and an update of the P6 Version 8.2 **Timescaled Logic Diagram** module.

It is a separate piece of Windows Client software that reads your Oracle Primavera database and allows a higher level of Gantt Chart customization than available from the Activities Window, in either Professional or Web Client. Version 16 added **Claim Digger** to Visualizer and named it **Schedule Comparison**.

Visualizer may be run from within P6 or from the Windows Start Menu and will create one of the following:

❖ **Timescaled Logic Diagrams** (TSLD),

❖ **Gantt Charts**, and

❖ Primavera P6 Version 16 moved **Schedule Comparison** to the **Visualizer** menu.

15 UNDERSTANDING P6 DATE FIELDS

Primavera has many more date fields for the current schedule than other products and some do not display as expected.

This section explains how these date fields calculate and indicates which to avoid.

After you understand these date fields, you should look again at the **Bar Timescale** options in the **Bars** form and it will be easier for you to understand how the bar formatting works.

To help understand the issues of some P6 date fields it is beneficial to understand how P6 calculates the Critical Path.

Basically, P6 calculates the forward and backward pass on all activities, including complete activities, using the **Remaining Duration** and the **Data Date**. This explains why completed activities have:

❖ **Early Dates** on the **Data Date**,

❖ **Late Dates** after the **Data Date** and

❖ It is possible to display a **Total Float** bar on a completed activity.

After you understand these date fields, you should look again at the **Bar Timescale** options in the **Bars** form and it will be easier for you to understand how the bar formatting works.

NOTE: The **Hint Help** is very useful to understand how fields are calculated.

15.1   Early Start and Early Finish

These are always the earliest dates that un-started activities or the incomplete portions of in progress activities may start or finish based on calendars, relationships and constraints.

❖ The **Early Start** of the completed activity A1010 is set to the **Data Date** date and time after the activity has commenced, not to the **Actual Start**, as in most other software.

❖ The **Early Finish** of the completed activity A1010 is set to the **Data Date** date and time when the activity is complete, not to the **Actual Finish**, as in most other software,

❖ The **Early Start** of an in progress activity is set to the **Activity Calendar** start after the activity has commenced, not to the **Actual Start**, as in most other software. This is more commonly known as the **Remaining Early Start**.

NOTE: Look carefully at the activity A1010 **Early Start** and **Early Finish** dates and then look at the **Actual Start** and **Actual Finish** of the bar; they are very different:

Activity ID	Activity Name	Early Start	Early Finish	September 2014				October 2014				November 201		
				25	01	08	15	22	29	06	13	20	27	03	10	17
A1010	Activity A	12-Oct-14 00	12-Oct-14 00													
A1020	Activity B	13-Oct-14 08	24-Oct-14 17													
A1030	Activity C	27-Oct-14 08	21-Nov-14 17													

The **Early Start** and **Early Finish** dates of completed activities and **Early Start** of in progress activities is not displayed in other software in this way and often leads to confusion when converting from other software.

NOTE: It is recommended that these are never displayed on an in progress schedule.

15.2 Late Start and Late Finish

❖ These are the latest dates that un-started activities or the incomplete portions of in progress activities may start or finish based on calendars, relationships, and constraints.

❖ The complete activity has the Late dates set the date that is equivalent to the latest point in time that the activity could be restarted.

❖ The **Late Start** of the in progress activity is actually the **Remaining Late Start**.

❖ The Total Float on the Complete Activity is "Null", but the default Layout shows a Float Bar.

❖ The Total Float bar finish is the same as the Late Finish and used to calculate Total Float.

NOTE: It is recommended these are never displayed on an in progress schedule.

Activity ID	Activity Name	Late Start	Late Finish	Total Float	September 2014					October 2014				November 2014			
					25	01	08	15	22	29	06	13	20	27	03	10	17	24
A1010	Activity A	20-Oct-14 08	20-Oct-14 08														
A1020	Activity B	20-Oct-14 08	31-Oct-14 17	5d													
A1030	Activity C	03-Nov-14 08	28-Nov-14 17	5d													

15.3 Actual Start and Actual Finish

These dates are manually applied, representing when an activity started or finished, and override constraints and relationships. These dates should be set in the past in relation to the **Data Date**.

Activity ID	Activity Name	Actual Start	Actual Finish	September 2014					October 2014				November 20		
				25	01	08	15	22	29	06	13	20	27	03	10	17
A1010	Activity A	01-Sep-14 08	26-Sep-14 17												
A1020	Activity B	29-Sep-14 08													
A1030	Activity C														

Actual dates should never change after they are assigned, but both the **Apply Actuals**, when activities are set to **Auto Compute Actuals**, and **Update Progress** functions may change Actual dates.

NOTE: Both these functions must be used with extreme caution.

15.4 Planned Dates Calculations

The **Planned Finish** is calculated from the **Planned Start** plus the **Original Duration**. The **Original Duration** is labeled **Planned Duration** in some **Industry Versions**. These fields are always linked, therefore:

❖ A change to the **Planned Start** will change the **Planned Finish** via the **Original Duration**,

❖ A change to the **Planned Finish** will change permanently the **Original Duration**, and

❖ A change to the **Original Duration** will change the **Planned Finish**.

In the pictures below the **Planned Dates** are displayed in the lower bar and when an activity has **NOT** started:

❖ The **Planned** dates **ARE** normally linked to the **Start** and **Finish**.
NOTE: Planned dates become unlinked from the **Start** and **Finish** when **Link Budget and At Completion for started activities is unchecked**. It is recommended you do not do this.

Start	Planned Start	Finish	Planned Finish	Jan 09							Jan 16					
				M	T	W	T	F	S	S	M	T	W	T	F	S
09-Jan-23 08	09-Jan-23 08	20-Jan-23 16	20-Jan-23 16													

When an activity is marked as in progress:

❖ The **Planned Start** date remains unchanged when an **Actual Start** date is set, even when it is different to the **Planned Start**.

❖ Therefore, the **Planned Start** remains the same as the **Start Date** before the **Actual Start** was set.

Start	Planned Start	Finish	Planned Finish	Jan 02							Jan 09							Jan 16					
				S	M	T	W	T	F	S	S	M	T	W	T	F	S	S	M	T	W	T	F	S
02-Jan-23 08 A	09-Jan-23 08	13-Jan-23 16	20-Jan-23 16																					

❖ The **Planned Finish** is calculated from the **Planned Start** plus the **Original Duration**.

© *Eastwood Harris*

NOTE: At this point the **Planned Dates** may contain irrelevant information, as they may not represent the schedule was planned, last period or this period, or match a baseline. The effectively contain irrelevant information and **MUST NEVER EVER** be displayed.

When an activity is complete:

❖ The **Planned Dates** remain unlinked from all other date fields.

15.5   Planned Dates Issues

This is one of the most important paragraphs in this book and you must be certain that you understand the Planned Dates and how to avoid the issues associated with them.

In the situation where a schedule is in the process of being updated:

❖ Assume the **Data Date** has been moved to the new **Data Date** and the project scheduled,

❖ Now all un-started activities will have their Start and Finish dates in the future,

❖ At this point every activity that is marked in progress by assigning an **Actual Start** (which should be in the past in relation to the **Data Date**) will have **Planned Dates** that neither:

➢ Match the status of the activity before the activity was marked as Started, nor

➢ Match the status of the activity after the activity was marked as Started and possibly finished.

Thus, in this situation, which is very common, the **Planned Dates** are now holding irrelevant dates that should never be displayed or used for any purpose.

Unfortunately, the Planned Dates are used by default in several places and Database Administrators and users must be aware of where they are used and how to avoid displaying them.

❖ The Planned dates are displayed as the **Project Baseline** bars and **Primary User Baseline** bars when no baseline has been assigned. Therefore, you should never display a Baseline Bar or columns unless a baseline project has been created and assigned, otherwise the Baseline bar and columns may represent irrelevant data.

❖ These **Planned Dates** are used by the **Apply Actuals** function, when activities are set to **Auto Compute Actuals**, and the **Update Progress** function is used. Thus, **Actual Start** dates and **Early Finish** dates of in progress activities will be changed to the **Planned Date** values without warning. The pictures show before and after **Update Progress** has been applied and you will see that the Actual Start has been changed without warning.

Before applying **Update Progress**

Start	Planned Start	Finish	Planned Finish	Jan 02	Jan 09	Jan 16
02-Jan-23 08 A	09-Jan-23 08	13-Jan-23 16	20-Jan-23 16			

After applying **Update Progress**

Start	Planned Start	Finish	Planned Finish	Jan 02	Jan 09	Jan 16
09-Jan-23 08 A	09-Jan-23 08	20-Jan-23 16	20-Jan-23 16			

NOTE: Ensure you **NEVER EVER** use the **Update Progress** function on a schedule that has been progressed, otherwise **Actual Start** dates and **Early Finish** dates of in progress activities will be changed to the **Planned Date** values without warning.

Some schedulers run a **Global Change** to set the planned dates to the **Start** and **Finish** dates before running **Update Progress**, but this has two issues:

➢ You must remember to run the **Global Change** before running **Update Progress**, and

➢ This changes the value of the **Duration % Complete**.

❖ The **Planned Dates** from a baseline schedule will be displayed as the **Baseline Bars** when the **Admin, Admin Preferences...**, **Earned Value** tab is set to **Budget values with planned dates**. Thus, the Baseline Bars read from an in progress baseline schedule are very likely to display incorrect data.

❖ Ensure **Admin, Admin Preferences...**, **Earned Value** tab has this value set as **At Completion values with current dates** or **Budget Values with current dates**. When the schedule is not resourced or cost-loaded it does not matter which of these two you use.

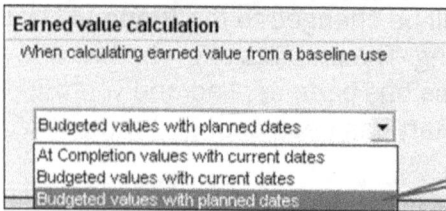

Earned value calculation

When calculating earned value from a baseline use

Budgeted values with planned dates ▾

At Completion values with current dates
Budgeted values with current dates
Budgeted values with planned dates

Never select
this option

15.6 How to use Update Progress on an In Progress Schedule

If you wish to use the **Update Progress** function on an in progress schedule, then this may be achieved by running a **Global Change** first that sets:

❖ The **Planned Start** date to the **Start,** and then

❖ Sets the **Planned Finish** date to the Finish by setting the **Original Duration** to the **At Completion Duration**.

This is a common process, but the author prefers not to use this process for two reasons:

❖ Firstly, the scheduler has to remember to run the **Global Change** before running **Update Progress**, and

❖ Secondly, the **Duration % Complete** will change when the **Original Duration** is changed by the Global Change. This may be desirable, but the **Original Duration** (the activity duration just before it was marked as started) is no longer the **Original Duration** after the **Global Change** is run.

15.7   Remaining Early Start and Finish

These are the earliest dates that the incomplete portions of un-started or in progress activities may start and finish.

❖ They are blank when an activity is complete.

❖ They may be edited in the same way as Planned Dates.

> ➢ When a **Remaining Early Start** is edited to a later than scheduled date, there is an option for constraining the **Remaining Early Start** with a **Start on or After Constraint**. If this is not set, then the activity will move forward to its original position when scheduling.

> ➢ When a **Remaining Early Finish** is edited, the **Remaining Duration** is also edited and the change is permanent. Scheduling does not take the schedule back to the original position.

Activity ID	Activity Name	Remaining Early Start	Remaining Early Finish	September 2014					October 2014				November 2014			
				25	01	08	15	22	29	06	13	20	27	03	10	17	24
A1010	Activity A																
A1020	Activity B	13-Oct-14 08	31-Oct-14 17														
A1030	Activity C	03-Nov-14 08	28-Nov-14 17														

15.8   Remaining Late Start and Finish

These are the latest dates that the incomplete portions of activities may start and finish.

❖ They are blank when an activity is complete and may not be edited,

❖ They may not be displayed as a bar,

❖ They are set to equal the **Late Dates**.

Activity ID	Activity Name	Remaining Late Start	Remaining Late Finish	September 2014					October 2014				November 2014				D
				25	01	08	15	22	29	06	13	20	27	03	10	17	24	01
A1000	Activity A																	
A1010	Activity B	13-Oct-14 08	07-Nov-14 17															
A1020	Activity C	10-Nov-14 08	05-Dec-14 17															

16 SCHEDULING OPTIONS

16.1 Setting the Schedule Options

When a project is rescheduled, there are some options available in the **Schedule Options** form, which is opened by selecting **Tools**, **Schedule...**, **Options** which control how the schedule calculates:

Use scheduling options from

EC00515 - City Center Office Building Addition

General | Advanced

☐ Ignore relationships to and from other projects

☐ Make open-ended activities critical

☑ Use Expected Finish Dates

☐ Schedule automatically when a change affects date

☐ Level resources during scheduling

☐ Recalculate assignment costs after scheduling

When scheduling progressed activities use

⦿ Retained Logic   ○ Progress Override

Calculate start-to-start lag from

⦿ Early Start   ○ Actual Start

Define critical activities as

⦿ Total Float less than or equal to

    0h

○ Longest Path

Calculate float based on finish date of

⦿ Each project   ○ Opened project

Compute Total Float as

Finish Float = Late Finish - Early Finish

Calendar for scheduling Relationship Lag

Successor Activity Calendar

> In P6 Version 20 and later, select the project **Scheduling Options** which will be assigned permanently and used to calculate all open projects

> Select **Make open-ended activities critical** when multiple critical paths are required

> Select **Longest Path** when using multiple calendars are assigned to activities

❖ Pressing the ▷ Default will set the options back to the P6 defaults, it does not save your options as the default.

© *Eastwood Harris*

❖ The **P6 Default** options are good but some need to be changed to suit specific situations.

➤ **Make open-ended activities critical** if you want a **multiple critical path** schedule, and

➤ **Define critical activities** as, **Longest Path** is recommended for projects with multiple calendars.

❖ When more than one project has been opened then it is important to understand how the **Default Project** and **Use scheduling options from** functions operate.

❖ **Use scheduling options from** was introduced in P6 Version 20 and overrides the project selected in **Project, Default Project.**

❖ In summary:

➤ The selected project **Scheduling Options** or **Default Project** in Version 19 and earlier are used to schedule all the opened projects, but

➤ More importantly all the project's **Scheduling Options** of all opened projects are changed permanently to the **Default Project Scheduling Options** or **Default Project** in Version 19 and earlier and they will not calculate the same way again.

❖ To prevent calculation issues, it is best to assign all projects opened together the same **Scheduling Options**.

❖ If you export a schedule to another database it is prudent to send a copy of the project **Scheduling Options** so they may be checked when imported into another database, especially if the schedule is to be opened with other projects.

16.2 Ignore relationships to and from other projects

Check this in the **Tools**, **Schedule...**, **Options** form to ignore relationships with other projects that are currently not open.

These relationships may be created between two projects when:

❖ Two or more projects are opened together, or

❖ When assigning a relationship, another project may be opened from the **Assign Predecessors** or **Assign Successors** forms and a relationship is created to an activity in another project that is not open.

This option will also ignore **External Dates**, which are the **External Early Start** and **External Late Finish** dates.

External Dates are constraints created when a project is exported from Primavera Contractor and/or another P6 database and imported into P6. They act like Early Start and Late Finish Constraints and are used to represent the relationships that would have originally provided the Early Start and Late Finish dates to the Critical Path calculations of the imported schedule.

NOTE: When you import a schedule always check for **External Dates**.

16.3 Make open-ended activities critical

An open-ended activity is an activity without a successor and normally has float to the end of the project. Checking the box makes these activities critical with zero total float when they do not have a successor.

This allows the user to display **multiple critical paths** in one project without the use of constraints and is useful, should you wish to see the individual critical paths for each area of a project. In order for this function to work the last activity in each chain of events must not have a successor:

❖ Open-ends Not Critical:

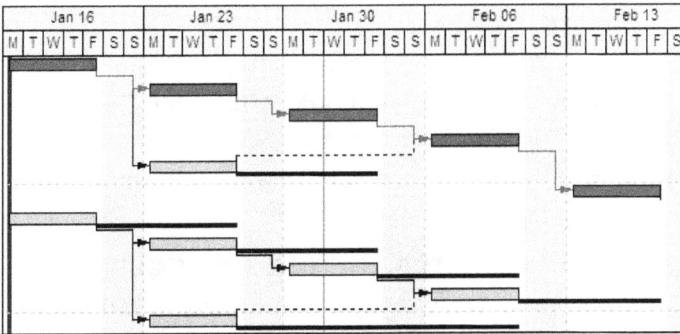

❖ Open-ends Critical:

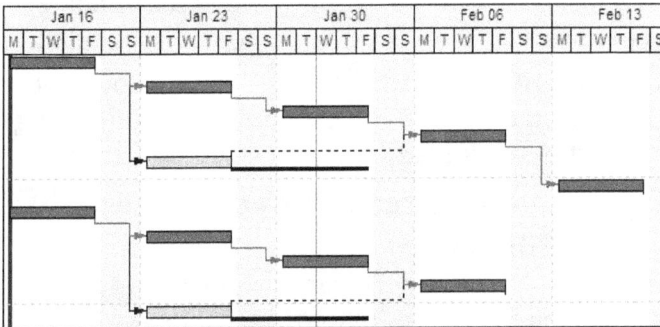

16.4 Use Expected Finish Dates

The intention of this option is for people using timesheets to be able to set an **Expected Finish** constraint for an activity.

Once an **Expected Finish** date is set, then the software calculates the **Remaining Duration** from:

❖ The **Early Start** when an activity has not started, or

❖ The **Data Date** when an activity has started, or

❖ A **Resume date** if a **Suspend** and **Resume** date has been set.

Therefore, **Expected Finish** dates may be assigned from the Timesheets module and this option allows the project manager to ignore these dates submitted with the timesheets.

16.5   Schedule automatically when a change affects dates

This is similar to automatic recalculation in other products and this recalculates the schedule when data that affects the timing of the schedule is changed.

NOTE: This may slow down your work significantly; this option is usually left off.

16.6   Level resources during scheduling

Leveling a schedule will delay activities until resources become available.

This is a form of resource optimization and this option levels the project resources each time it is scheduled.

NOTE: This is **NOT** recommended as it slows down the schedule calculation and the schedule will often change each time it is scheduled.

16.7   Recalculate resource costs after scheduling

Resource Unit Rates may be set to change over time in the **Units & Prices** tab of the **Resources Window**:

This option recalculates a resource cost when a resource is scheduled into a different cost rates time bracket.

16.8   When scheduling progressed activities use

"Out of Sequence Progress" occurs when an activity starts before a predecessor defined by a relationship has finished. Therefore, the relationships have not been acknowledged and the successor activity has started out of sequence.

There are three options in P6 for calculating the finish date of a successor when the successor activity has started before the predecessor activity is finished:

❖ **Retained Logic**
❖ **Progress Override**
❖ **Actual Dates**

The picture below represents the status of the activities before updating the schedule:

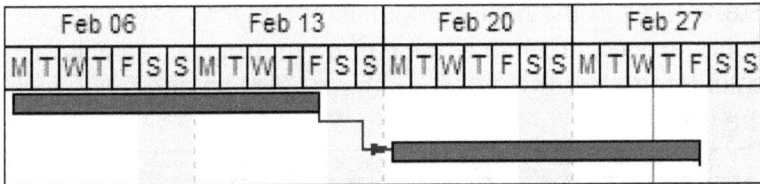

Retained Logic.

In the example following, the relationship is maintained between the predecessor and successor for the unworked portion of the activity (the Remaining Duration) and continued after the predecessor has finished.

The relationship forms part of the critical path and the predecessor has no float.

NOTE: This is the recommended option as a more conservative schedule is produced and any relationships may be changed as required:

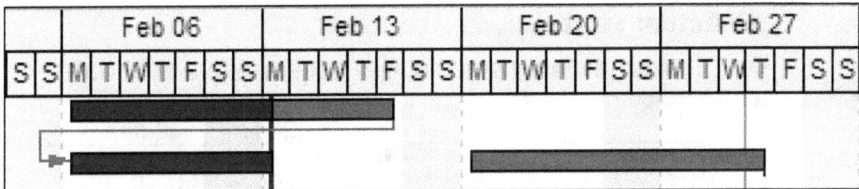

Progress Override.

In the example following, the Finish-to-Start relationship between the predecessor and successor is disregarded, and the unworked portion of the activity (the Remaining Duration) continues before the predecessor has finished.

NOTE: The relationship is not a driving relationship and **DOES NOT** form part of the critical path in the example below, and the predecessor has float:

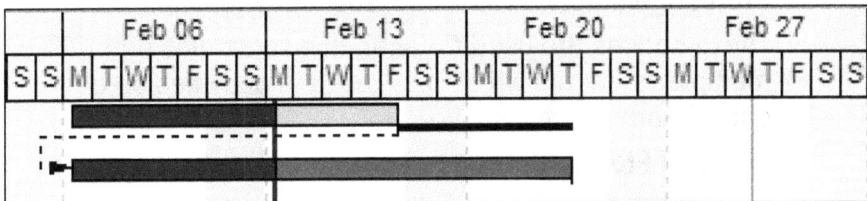

Actual Dates.
This function operates when there is an activity with Actual Start Dates in the future, which is not logical. With this option the remaining duration of an in progress activity is calculated after the activity with actual start and finish in the future:

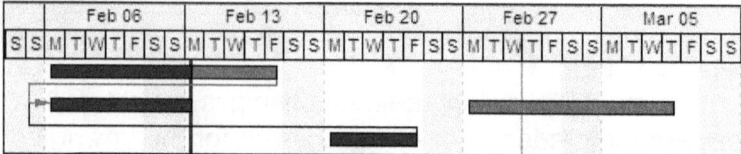

When there are no Actual Dates in the future this option calculates as Retained Logic.

16.9   Calculate start-to-start lag from

The successor of an activity with a Start-to-Start and positive lag would start after the lag has expired. When the predecessor commences out of sequence the lag may be calculated from the predecessor calculated **Early Start** or the **Actual Start**.

❖ The **Actual Start** gives a less conservative schedule:

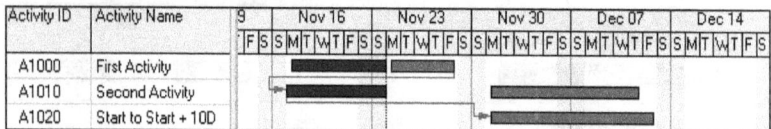

❖ The **Early Start** gives a more conservative schedule:

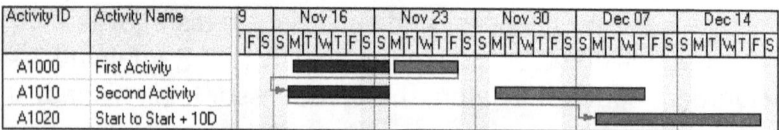

16.10 Define critical activities as

Critical Activities Definition criteria is defined in the **Projects Window**, **Project Details**, **Settings** tab.

These options are used for analyzing schedules that utilize multiple calendars which may result in activities on the critical path possessing float.

❖ **Total Float less than or equal to** – Activities may be marked as critical and with a chosen float value.

Sometimes a small positive value is used to isolate the near critical activities on schedules or displaying the full critical path on multiple calendar schedules.

❖ **Longest Path** – Oracle define in their Help File as "*In a multicalendar project, the longest path is calculated by identifying the activities that have an early finish equal to the latest calculated early finish for the project and tracing all driving relationships for those activities back to the project start date.*"

❖ In the example below the Total Float has been set to **Total Float less than or equal to zero** and the critical path has disappeared:

Calendar	Critical	Total Float	Aug 22
			Mon · Tue · Wed · Thr · Fri · Sat · Sun · Mon · Tue
7 Day/Week	☐	2d	
6 Day/Week	☐	1d	
5 Day/Week	☑	0d	

❖ When the **Total Float** is then set to less than or equal to 1 day, it results in the picture below:

Calendar	Critical	Total Float	Aug 22
			Mon · Tue · Wed · Thr · Fri · Sat · Sun · Mon · Tue
7 Day/Week	☐	2d	
6 Day/Week	☑	1d	
5 Day/Week	☑	0d	

❖ When the **Total Float** is then set to **Longest Path**, it results in the picture below:

Calendar	Critical	Total Float	Aug 22
			Mon · Tue · Wed · Thr · Fri · Sat · Sun · Mon · Tue
7 Day/Week	☑	2d	
6 Day/Week	☑	1d	
5 Day/Week	☑	0d	

NOTE: Longest Path is recommended for projects with multiple calendars.

16.11 Calculate float based on finish date of

This is a new function to Version 6.2. When more than one project is opened, the Total Float may be calculated based on each individual project, or the longest project:

❖ **Each project** – used when each project's critical path is required:

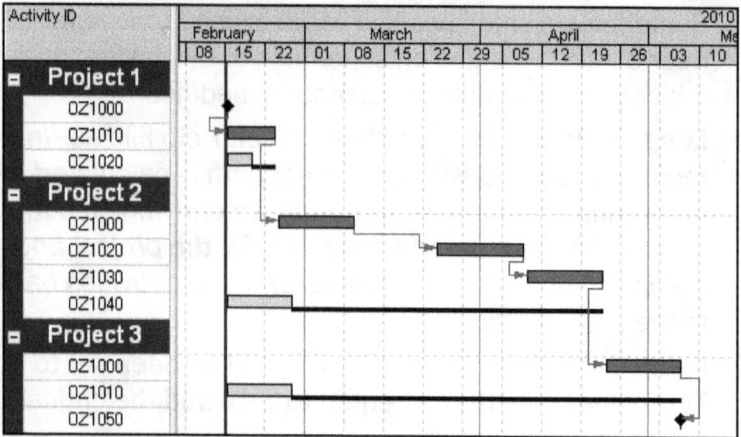

❖ **Opened projects** – used when all the P6 projects are related and float is required to be based on the longest project:

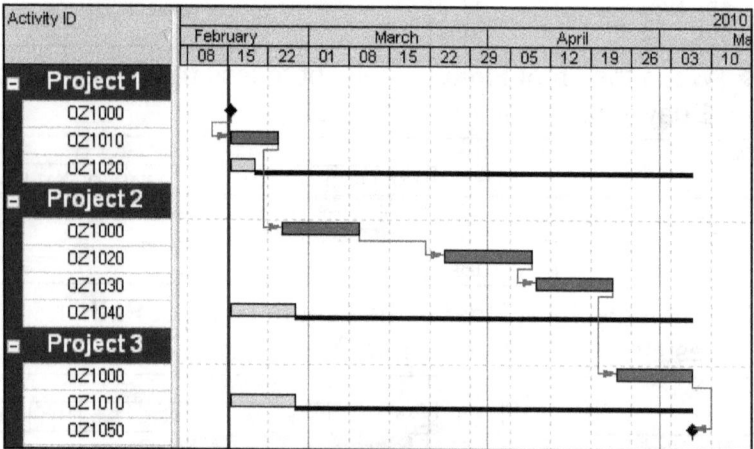

NOTE: The issue with the **Opened projects** setting is that, depending on what projects are opened, the Total Float will calculate differently and is therefore not recommended.

16.12 Compute Total Float as

There are three options for the calculation of the Float value displayed in the Total Float column of WBS and LOE activities only and is not an important setting.

16.13 Calendar for scheduling Relationship Lag

❖ There are four calendar options for the calculation of the lag for all activities:

1. **Predecessor Activity Calendar** is the default, the example below has a 40-hour lag, or

Calendar				Jun 01						
	Sun	Mon	Tue	Wed	Thr	Fri	Sat	Sun	Mon	Tue
24 Hours/Day 7 Days/Week										
5 Day/Week										

2. **Successor Activity Calendar.** note the change in the successor start, or

Calendar				Jun 01						
	Sun	Mon	Tue	Wed	Thr	Fri	Sat	Sun	Mon	Tue
24 Hours/Day 7 Days/Week										
5 Day/Week										

3. **24-Hour,** or

4. **Project Default Calendar.**

Microsoft Project 2003 to 2019 uses the successor calendar for lag calculations. Microsoft Project also has the option of an Elapsed lag duration and % Duration lag.

Elecosoft (Asta) Powerproject does not assign lags to the relationship but a lag is assigned to the predecessor or successor activity thus allowing an unlimited number of relationships between two activities and a partial Critical Path, plus it also has a % Duration lag.

17   SETTING THE BASELINE

Setting the Baseline makes a complete copy of a project, including relationships, notebook entries and codes. You are then able to compare the current project's progress against the baseline.

NOTE: A big issue with P6 is that not all the data from a Baseline is readable from the current project.

17.1   Typical Baseline types

There are typically three types of Baselines:

❖ Management, for recording a contract schedule

❖ Last Period Status, for comparing one period's status with another, and

❖ Claims Analysis, to record the status before applying a delay or acceleration to a schedule:

17.2   Number of Baseline Allowed

An unlimited number may be saved, but there is still a restriction of copying a maximum of 50 Baselines when copying a project.

The number that may be saved or copied is set in the **Admin Preferences**, **Data Limits**.

17.3   Setting a Baseline

Setting a Baseline takes place in two steps.

❖ Baselines are created, deleted, restored and updated in the **Project**, **Maintain Baselines** form,

❖ The baselines are assigned from the **Project**, **Assign Baselines...** form.

A Baseline may not be opened unless it is **Restored** and at that point it is not a baseline.

17.4 Diplaying Baseline Data

Baseline data may be displayed:

❖ Up to four Baselines may be assigned,

❖ Baseline data may be viewed from the current schedule in columns or as bars,

❖ More data may be seen from the **Project Baseline** and **Primary User Baseline** than from the **Secondary User Baseline** and **Tertiary User Baseline**.

17.5 Difference between the Project Baseline and a User Baseline

Often two baselines are set, one representing the current agreed contract schedule and one representing the last period progress, allowing users to see the variance in the last period and from the contract agreed schedule.

❖ When a **Project Baseline** is set, any user that opens a project will see the same baseline.

❖ When a **User Baseline** is set, then other users will not see this baseline.

NOTE: If more than one user and **User Baselines** are being used, then there must be corporate procedures to ensure that each user sets the same user baseline.

17.6 Limitation on Viewing Baseline Data

The following types of data may **NOT** be read from a **Project Baseline** and **Primary User Baseline**:

❖ Late Dates and Late resource data,

❖ Total and Free Float,

❖ Relationships,

❖ Constraints,

❖ % Complete, Actual and Remaining Durations,

❖ Individual resources and costs and units, these are read at Resource Type, thus if there are three Labour Resources, then the Baseline will display the total of all the Labour Resources,

❖ Material Resource and Expense Units.

The following types of data may **NOT** be read from a **Secondary User Baseline** and **Tertiary User Baseline**:

❖ Activity Durations,

❖ Any Resource or Expense data.

17.7 Understanding the <Current Project> in the Assign Baseline form

When:

❖ Either a **Project Baseline** bar or **Primary User Baseline** bar is being displayed, and

❖ A Baseline has NOT been set,

Then a baseline bar will read the **Current Schedule Planned Dates**, which may hold irrelevant data and are not baseline data.

NOTE: When you display a **Project Baseline** bar or **Primary User Baseline** bar, make sure you have the appropriate baseline set. Otherwise you will be displaying the **Planned Dates** from the **Current Schedule** and if they have progress and they have been marked as **Started**, then they may hold irrelevant information.

Furthermore, after setting a Baseline, you also need to ensure that you are not reading the Baseline Planned Dates by changing the default **Admin Preferences Earned Value** setting so it is not reading **Budget values with planned dates**.

18 UPDATING AN UNRESOURCED SCHEDULE

18.1 Updating Process

P6 is a very simple product to update compared to many of its competitors. The process to update a project is:

❖ Saving a **Baseline** schedule,

❖ Recording or marking-up progress at the **Data Date**,

❖ Update or Progress the schedule:

➢ **Completed** activities are assigned **Actual Start** and **Actual Finish** dates,

➢ In progress activities are assigned **Actual Start** dates, and the activity's **Remaining Durations** and the **Percent Completes** are adjusted,

➢ Adjustments are made to un-started work based on the productivity to-date, and

❖ Project scope changes should be added as new activities.

❖ **Scheduling** the project, at the same time moving the **Data Date** to the new **Data Date** and recalculating all the activities dates. The **Data Date** may also be moved before updating the activities from the **Project Window**, **Dates** tab.

❖ Comparing and reporting actual progress against planned progress and revising the plan and schedule, if required.

Comparing the status of an activity against more than one baseline is useful; for example:

❖ The original plan could be represented as one of the Baselines, to see the slippage against the original plan.

❖ Last Period, which could be another Baseline, to see the changes since the last update.

18.2 Calculation of Durations of an In progress Activity

The Primavera program has many duration fields, and we will discuss four duration fields below:

❖ An activity **Original Duration (Planned Duration)** in some Industry Versions and in the Web tool) is the duration from the **Early Start** to the **Early Finish** calculated over the **Activity Calendar** and is calculated when an activity has not yet started. When an **Actual Start** is entered, this duration is no longer recalculated or directly used for scheduling, but may be edited.

❖ The **Actual Duration** is the activity's worked duration and is either the duration from:

  ➢ The **Actual Start** to the **Data Date** of an in progress activity, or

  ➢ The **Actual Start** to the **Suspend Date** of a suspended in progress activity, or

  ➢ The **Actual Start to** the **Actual Finish** of a **Completed** activity.

❖ The **Remaining Duration** is the unworked duration of an **In progress** activity and is the duration from the **Data Date** or **Resume Date** to the **Early Finish** date of an activity.

❖ The **At Completion Duration = Actual Duration + Remaining Duration**. Before an activity has started, the **Actual Duration** is zero and the **Remaining Duration** equals the **Original Duration**.

❖ The **Original Duration** is linked to the **Remaining Duration** when an activity is un-started and the **Link Budget and At Completion for not started activities** box in the **Calculations** tab of the **Projects Window** is checked.

❖ The Remaining bar is based on the **Remaining Duration**, and the **Remaining Duration** may commence a period of time after the **Data Date**, so there is often a gap between the **Data Date** and the **Remaining Start** of an in progress activity.

NOTE: The in-built proportional link between **Original Duration**, **Actual Duration**, **Remaining Duration**, and **% Complete** that exists in Microsoft Project does not exist in Primavera.

18.3   Default Percent Complete

The Activity % Complete is displayed on the **% Complete Bar**, and may be linked to only one of the following three % Complete fields:

❖ **Physical % Complete**

❖ **Duration %Complete**

❖ **Units % Complete**

Each new activity is assigned the project default **Percent Complete Type** and then this may be edited for each activity as required.

The **Default % Complete Type** for each new activity in each project is assigned in the **Defaults** tab of the **Details** form in the **Project Window**:

❖ **Duration % Complete** – This field is calculated from the proportion of the **Original Duration** and the **Remaining Duration** and they are linked, and a change to one value will change the other. When the **Remaining Duration** is set to greater that the **Original Duration** this percent complete is always zero.

❖ **Physical % Complete** – This field enables the user to enter the percent complete of an activity and this value is independent of the activity durations. This is similar to the way P3 and SureTrak calculate the % Complete when the **Link Remaining Duration and Percent Complete** option is NOT selected.

❖ **Units % Complete** – This is where the percent complete is calculated from the resources Actual and Remaining Units. A change to one value will change the other and when more than one resource is assigned, then all the Actual Units for all resources will be changed proportionally. This is similar to the Microsoft Project % Work Complete.

NOTE: The **Units % Complete** is calculated from the value of all the Labor and Non-Labor Resources, so be careful when more than on type of resource is assigned to an activity. For example, the software could be adding concrete volumes with labor hours and excavator hours.

18.4 Activity % Complete

The **Activity % Complete** field is linked to the **% Complete Type** field assigned to an activity in the **General** tab of the **Details** form in the **Activities Window**, or the **% Complete Type** column:

The **Activity % Complete** is also linked the **% Complete Bar** and this value is represented on the **% Complete** Bar.

Percent Complete Type	Original Duration	Remaining Duration	Activity % Complete	Duration % Complete	Physical % Complete	Units % Complete	Actual Labor Units	At Completion Labor Units	Oct 24	Oct 31	N
% Co...	10d	6d		40%		33.33%	5	15			
Duration	10d	6d	40%	40%	5%	0%	0	0			
Physical	10d	6d	12%	40%	12%	0%	0	0			
Units	10d	5d	33.33%	50%	5%	33.33%	5	15			

18.5 Understanding the Current Data Date

The Primavera **Current Data Date** is displayed as a vertical line on the schedule; this **Data Date** vertical line may be formatted in the **Bar Chart Options** form.

In P6 the function of the **Current Data Date** is to:

❖ Separate the completed parts of activities from incomplete parts of activities.

❖ Calculate or record all costs and hours to-date before the **Current Data Date**, and to forecast costs and hours to go after the **Current Data Date**.

❖ Calculate the **Finish Date** of an in progress activity from the **Current Data Date** plus the **Remaining Duration** over the **Activity Calendar**, when the **Suspend** and **Resume** function has not been used.

18.6 Updating the Schedule

The schedule may be updated using the following methods:

❖ Using the fields in the **Status** tab of the **Details** form in the lower pane, or

❖ Displaying the appropriate tracking columns by:

  ➢ Creating your own layout, or

  ➢ Inserting the required columns in an existing layout.

18.6.1 Updating Activities Using the Status Tab of the Details Form

Ensure you are showing the time and then open the **Status** tab:

Updating a Complete activity:

❖ Check the **Started** box and enter the actual **Start Date and Time** if different from the displayed date.

❖ Check the **Finished** box and enter the actual **Finish Date and Time** if different from the displayed date.

Updating an In progress activity:

❖ Check the **Started** box and enter the actual **Start Date and Time** if different from the displayed date.

❖ When the **Percent Complete Type** is **% Duration** the **% Duration Complete** and **Remaining Duration** are linked, then when:

> ➤ The **Remaining Duration** is edited and the **% Complete** is calculated, or

> ➤ The **% Complete** is entered and the software calculates the **Remaining Duration**, or

> ➤ A **Remaining Duration** greater than the **Original Duration** may be entered and the **% Duration** will remain at zero, until the **Remaining Duration** is less than the **Original Duration**.

Irrespective of the method used to calculate the **Remaining Duration**, after the schedule is recalculated the end date of the activity is calculated from the **Current Data Date** plus the **Remaining Duration** over the **Activity Calendar**.

NOTE: Be careful that the **% Duration Complete** does not change the **Remaining Duration** to a non-round day and that activity then finishes halfway through a day. This results in all the successor activities starting and finish in the middle of the day.

Updating an Un-started activity:

❖ The **Original Duration**, **Relationships** and **Constraints** of an un-started activity should be reviewed.

© *Eastwood Harris*

18.6.2 Updating Activities Using Columns

An efficient method of updating activities is by displaying the data in columns. This may be achieved by:

❖ Inserting the required columns in an existing layout, or better still:

❖ Creating a Layout with the required columns and updating the schedule using these columns.

18.7 Progress Spotlight and Update Progress

18.7.1 Progress Spotlight

Primavera Version 5.0 introduced a new function for highlighting the activities that should have progressed in the update period.

The user then has the option of selecting some or all of the activities that should be updated and updating them using the **Update Progress** function as if they progressed exactly as they were Planned.

The **Spotlight** may be moved to reflect the new **Data Date** by either:

❖ Dragging the **Data Date**, or

❖ Using the **Tools** toolbar, **Spotlight** icon 🔲 which highlights all activities that should have progressed in one minor time period of the timescale settings.

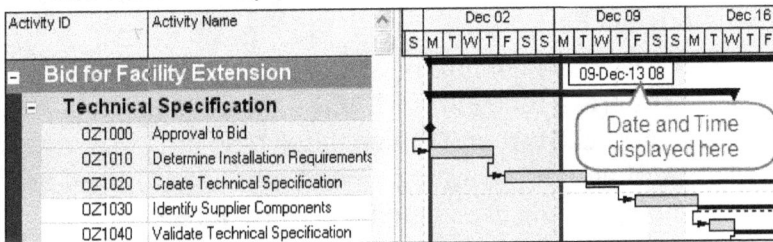

NOTE: Activities highlighted with a filter do not show in a **Print Preview**, but activities highlighted with **Progress Spotlight** will display in the **Print Preview**.

18.7.2   Update Progress

To update a schedule using the **Update Progress** form, select **Tools**, **Update Progress**:

Update Progress		×
Earliest Data Date	01-May-11	Apply
New Data Date	01-Jan-24 08:00 ...	Cancel
When using progress spotlight update progress for		Help
⦿ All highlighted activities		
○ Selected activities only		
When actuals are applied, calculate activity remaining durations:		
○ Based on activity duration type		
⦿ Always recalculate		

Unlike Microsoft Project the Primavera **Update Progress** facility uses the **Planned Start** and **Planned Finish** dates (not the **Early Start** and **Early Finish**) for setting the **Actual Start** and **Actual Finish** dates of in progress activities.

Thus, when the **Planned Dates** of an in progress activity are different from the **Actual Start** and **Early Finish** dates and the activity is Automatically updated to be complete, then both these dates are set to the **Planned** dates and the **Actual Start** may be changed and the **Actual Finish** not set to the original **Early Finish**.

NOTE: As discussed before it is recommended that the **Update Progress** function is **NEVER** used because it changes **Actual Start** and **Finish** dates of in progress activities without any warning. On the other hand, if you are using **Apply Actuals** or wish to use the **Update Progress** function then you should run a Global Change to set the **Planned** dates to the **Start** and **Finish**, which will prevent the **Actual Start** and **Finish** of an in-progress activity changing.

18.8   Suspend and Resume

The Primavera Version 5.0 Suspend and Resume function enables the work to be suspended and the activity resumed at a later date. Open the **Activity Details** form **Status** tab and enter the **Suspend** and **Resume** dates. This function enables only one break in an activity.

The following example shows an activity with a **Suspend** date and **Resume** date set:

❖ This feature works when an activity has commenced and normally the **Suspend** date is in the past and the **Resume** date in the future.

❖ The activity must have an **Actual start** date before you can record a **Suspend** date.

❖ Only **Resource Dependent** and **Task Dependent** activities may be suspended and resumed.

❖ The suspended period is not calculated as part of the activity duration and resources are not scheduled in this period.

NOTE: The Suspend and Resume time may be set at the incorrect time of the day. The author has found that the Suspend is usually set at the start of the day and Resume usually at the end of the day; therefore, the defaults for both are illogical. Therefore, you **SHOULD ALWAYS** display the time when setting Suspend and Resume dates to ensure that they are correct.

18.9 Scheduling the Project

At any time, but usually after some or all the activities have been updated, the project is scheduled:

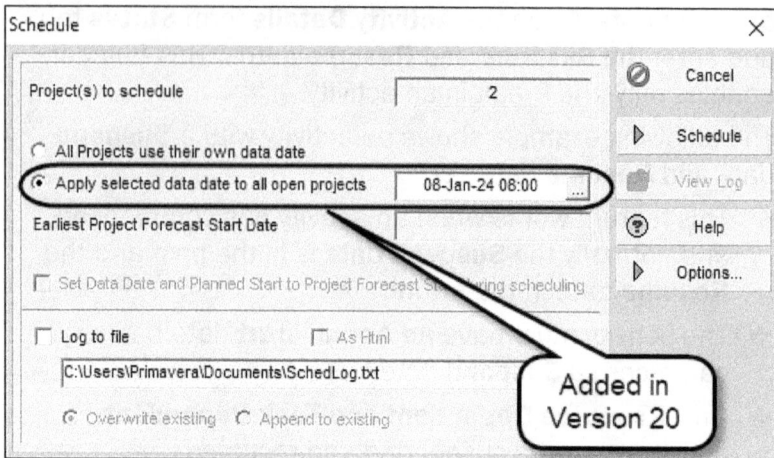

❖ Open the **Schedule** form:

> ➢ Select **Tools**, **Schedule**..., or

> ➢ Press the **F9** key, or

> ➢ Click on the 🕐 icon.

❖ Select the revised **Current Data Date and Time** from the box and click the ▷ Schedule icon.

❖ The software will recalculate all the early finish dates from the remaining durations and the new **Current Data Date**, taking into account the relationships and the **Schedule Options**.

❖ Primavera Version 20 added an additional option:

> ➢ **All Projects use their own data dates** which is how Version 19 and earlier versions calculated, or

> ➢ **Apply selected data date to all open projects** for Version 20 and later and all projects will have their **Data Date** permanently changed to the selected date.

18.10 Comparing Progress with Baseline

There will normally be changes to the schedule dates and more often than not there are delays. The full extent of the change is not apparent without having a Baseline bar to compare with the updated schedule.

To display one or more of the **Baseline Bars** in the **Bar Chart** you must open the **Bars** form and check the **Display** box of one or more baseline bars.

Activity ID	Activity Name	Variance - BL Project Finish Date	Dec 01	Dec 08	Dec 15	Dec 2
Bid for Facility Extension		0d				
Technical Specification		-4d				
OZ1000	Approval to Bid	-1d				
OZ1010	Determine Installation Requirements	0d				
OZ1020	Create Technical Specification	-4d				
OZ1030	Identify Supplier Components	-4d				
OZ1040	Validate Technical Specification	-4d				

If you want to see the Start and Finish Date variances, they are available by displaying the **Variance – BL Project Start Date**, **Variance–BL Finish Date**, **Variance– BL1 Start Date**, and **Variance– BL1 Finish Date** columns.

NOTE: Variance columns for **Secondary** and **Tertiary Baseline Dates** are not standard columns but could be calculated with a **Global Change**.

NOTE: As discussed earlier in this chapter, when you display a **Project Baseline** bar or **Primary User Baseline** bar, make sure you have the appropriate baseline set. Otherwise you will be displaying the **Planned Dates** from the **Current Schedule** and if they have progress and they have been marked as **Started**, then they may hold irrelevant information.

Furthermore, after setting a Baseline you also need to ensure that you are not reading the Baseline Planned Dates, by changing the default **Admin Preferences Earned Value** setting so it is not reading **Budget values with planned dates**.

18.11 Progress Line Display on the Gantt Chart

A progress line displays how far ahead or behind activities are in relation to the Baseline. Either the Project Baseline or the **Primary User Baseline** may be used and there are four options:

❖ Difference between the **Baseline Start Date** and **Activity Start Date**,

❖ Difference between the **Baseline Finish Date** and **Activity Finish Date**,

❖ Connecting the progress points based on the **Activity % Complet**e,

❖ Connecting the progress points based on the **Activity Remaining Duration**.

There are several components to displaying a Progress Line:

❖ Firstly, the progress line is formatted using the **View, Bar,** ▢ Options... form, **Progress Line** tab, which may also be opened by right-clicking in the Gantt Chart area:

❖ Selecting **View, Progress Line** to hide or display the **Progress Line**.

❖ If you use either of the options of **Percent Complete** or **Remaining Duration**, then you must display the appropriate Baseline Bar that has been selected as the **Baseline to use for calculating Progress Line:**

❖ The picture below shows the option of **Percent Complete**:

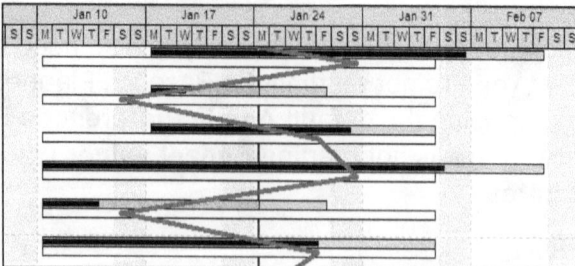

18.12 Comparing Schedules

After updating a schedule or receiving a schedule from a contractor and you wish to compare two schedules, then there are several methods that you may wish to consider:

❖ Setting one project as a baseline will allow limited comparison and is restricted to date, duration, cost and unit changes.

❖ Some people copy and paste the two projects into Excel and use VLookup formulae to compare the two projects.

❖ The **Schedule Comparison** utility allows you to select and compare two projects and almost all data changes. This is run from **Tools, Schedule Comparison**.

❖ There are many third party tools listed on **www.primavera.com.au/third_party_software.html** that you may wish to look at including:

  ➢ Zümmer Analysis
  ➢ Schedule Analyzer
  ➢ Change Inspector.

19 CREATING ROLES AND RESOURCES

19.1 Understanding Roles and Resources

Traditionally, planning and scheduling software defines a **Resource** as something or someone that is required to complete the activity and sometimes has limited availability. This includes people or groups of people, materials, equipment and money.

Primavera is able to assign Costs, a Calendar, one or more Roles and some personal information to a **Resource**.

Primavera has a function titled **Roles**. A Role is normally used at the planning stage of a project and represents a skill or position. Later, and before the activity begins, a Role would be filled by assigning a specific individual who would be defined as a Resource. Roles may be assigned to both Resources and Activities. A search by Role may be conducted on all the Resources when it is required to replace an Activity-Assigned Role with an individual from the Resource pool. Primavera allows rates to be assigned to Roles.

An evaluation at project or enterprise level in order to understand the long-term demand for resources may be made by a combination of Roles for long term planning and Resources for short term planning.

Roles are assigned to activities for long term planning and Resources represent individual people who are assigned to roles for short term planning and would represent activity assignment.

The light area to the right of the histogram below shows the unsatisfied demand of the roles titled **Unstaffed Remaining Units**. The satisfied demand from the resources is in the dark are to the left and is titled **Staffed Remaining Units**.

Filtered for activities with **PM Role assigned**

Unsatisfied demand of the roles titled **Unstaffed Remaining Units**

Satisfied demand from the resources titles **Staffed Remaining Units**

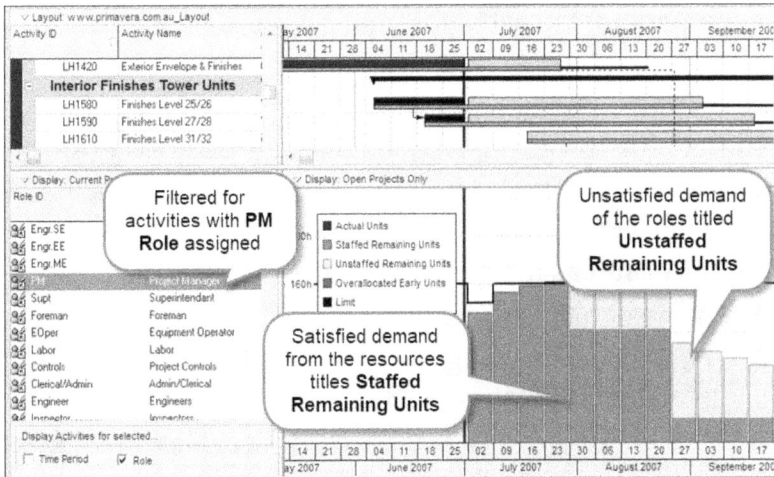

If you are a construction contractor and not assigning work to individual people, then you may consider using only resources as they have more functions than roles.

19.2   Creating Roles

Roles are created, edited, and deleted in a similar method as WBS Nodes.

To create, edit, or delete a **Role**, select **Enterprise**, **Roles...** to open the **Roles** form:

In the **General** tab each Role may be assigned a:

❖ **Role ID**, a unique code used to assign the Role to an Activity.

❖ **Role Name**, the name of the Skill or Trade.

❖ **Responsibilities**, where you may enter text, hyperlinks, and pictures about the Role Responsibility.

In the **Resources** tab each Role may be assigned:

❖ To one or more Resources.

❖ The Resource is assigned by default a **Proficiency** of "3 – Skilled" which may then be changed to any of the options shown in the list.

Proficiency
3 - Skilled ▼
1 - Master
2 - Expert
3 - Skilled
4 - Proficient
5 - Inexperienced

❖ The Resource may be assigned a **Primary Role** which would represent the task or job they would normally be assigned.

❖ The **Primary Role** also links to **Role** availability when the option in the **Edit**, **User Preferences...**, **Resource Analysis** tab, **Display the Role Limit based on** is set to **Calculate primary resources' limit**.

Primavera supports **Rates for Roles**. Up to 5 rates (the same number of rates as resources) may be assigned to roles which may be used for estimating and cash flow forecasting of projects before the actual resource completing the work is assigned to the activity.

❖ Click on the **Prices** tab to edit the Role Price/Unit, and

❖ The Role Rate Type is adopted from the Resource Rate Type set in the **Admin**, **Admin Preferences...**, **Rate Type** tab.

NOTE: Resources have always been able to vary their rate over time but Roles were not and this was one of the drivers to use resources and not roles on projects.

P6 Version 20 now allows Role rates to vary over time but not levelled.

TRICKS & TRAPS FOR **ORACLE PRIMAVERA** P6 PPM PROFESSIONAL

19.3   Creating Resources

19.3.1   Resources Window

To create, edit, or delete resources, open the **Resources Window**:

❖ Select **Enterprise**, **Resources...**, or

❖ Click on the 🖳 icon on the **Enterprise** toolbar,

Resources				
Resources	**Projects**			

Display: All Active Resources

Resource ID	Resource Name	Resource Type	Price / Unit	Unit of Measure
Machinist	Machinist	Labor	$23/h	
Construct Nonlabor	Material Resources	Material	$0/unit	
Piping Mat	Piping Material	Material	$3/cu yd	Cubic yards
mikeb	Mike Brown, CIO	Labor		
trents	Trent Smith, PMO Director	Labor		
johnm	John McDougal	Labor	$80/y	
frankl	Frank Lee	Labor	$75/h	

General	Codes	Details	Units & Prices	Roles	Notes

Resource ID	Resource Name
trents	Trent Smith, PMO Director

Employee ID	Title
123456789	Mr

E-Mail Address	Office Phone	
teamplayhelp@primavera.com	610-949-6800	☑ Active

❖ In the **Resources Window**:

➢ 🖳 indicates a **Resource** which is not assigned to an open project,

➢ 🖳 indicates a **Resource** assigned to an open project,

➢ 🗐 and 🗐 indicate an unassigned and assigned **Nonlabor Resource**, and

➢ 🗐 and 🗐 indicate an unassigned and assigned **Material Resource**.

143   © *Eastwood Harris*

19.3.2   Adding Resources

New Resources are added and deleted in a similar method to adding Activities in the **Activities Window**. Use the **Insert** key, right-click and select **Add**, or use the Add ⊞ icon on the **Edit** toolbar.

19.3.3   General Tab

The fields in this tab are self-explanatory:

❖ The **Resource ID** has to be unique within a database and a **Resource Name** is mandatory,

❖ When the **Active** box is unchecked, the Resource is inactive and indicates that the resource is not available. When assigning Resources to Activities there is a filter to display only active Resources.

General	Codes	Details	Units & Prices	Roles	Notes

Resource ID	Resource Name
ARL	Angela Lowe

Employee ID	Title
12345678	Ms

E-Mail Address	Office Phone	
angela.lowe@eh.com.au	03 9846 7700	☑ Active

19.3.4   Codes Tab

Resource Codes are assigned to Resources allowing additional facilities to sort and report on them in the **Resource Usage Spreadsheet** and **Resources Window**.

❖ **Resource Codes** may be defined in the **Resource Code Definition** form, which is opened by selecting **Enterprise, Resource Codes...** and clicking on the
🖻 Modify... icon.

❖ **Resource Codes** may then be selected in a layout to sort and group Resources for fields, such as location or employment type.

General	Codes	Details	Units & Prices	Roles	Notes

Resource Code	Code Value	Code Description
Clearance	S	Secret
Classification	FTE	Full Time Employee
Office Location	EUR.LON	London

19.3.5 Details Tab

Resource Types

There are three types of Resources:

❖ **Labor**, intended for people

❖ **Nonlabor**, intended for equipment

❖ **Material**, intended for materials/supplies.

Material Resources

May be leveled and have the following differences from other resources:

❖ They may be assigned a **Unit of Measure**, which is created in the **Admin, Admin Categories..., Unit of Measure** tab. This is not available to Labor and Nonlabor resources.

❖ They may not be assigned a Role.

❖ They may not log Overtime.

Material resources do not display units (quantities) in the **Activities Window**, unlike many other products like Elecosoft (Asta) Powerproject. These values may be displayed in other views such as the **Resource Assignments Window** and in reports.

Currency

An alternate **Currency** may be associated with a resource. This will not affect how the Resource Unit Rates costs are entered but provides a further tagging mechanism for sorting and reporting. The costs are stored in the default currency but are displayed using the conversion rate in the currency selected for the resource.

Overtime

A **Labor Resource** may be allowed to record **Overtime** in the **Primavera timesheet** system when the **Overtime Allowed** box is checked and the costs derived from the **Unit Rates** are multiplied by the **Overtime** Factor.

Calendar

The Resource is assigned a **Global** or a **Shared Resource Calendar** in this form.

❖ A Shared Resource Calendar may be created and assigned to more than one Resource. This topic is covered in more detail in the next section of this chapter.

❖ Click on the [Create Personal Calendar] icon to create a Personal Calendar for this resource.

NOTE: The **Resource** calendar is used to display the resource limits, irrespective of the calendar assigned to an activity.

Default Units/Time

The **Default Units/Time** is the value that a resource adopts when it is first assigned to an activity. In a similar way to Microsoft Project, the **Units per Time Period** may be displayed as a **Percentage** or in **Units/Time**.

❖ Select **Edit**, **User Preferences...**, Time **Units** tab and select the preferred display from the **Units/Time Format** section:

Units/Time Format	Default Units / Time
Resource Units/Time can be shown as a percentage or as units per duration	50%
○ Show as a percentage (50%)	Default Units / Time
⦿ Show as units/duration (4h/d)	4.00h/d

For example, you may have a fleet of 12 trucks and you usually assign four trucks to each loader. In this situation you would assign the **Default Units/Time** as 400%, or 4 d/d, or 32h/d if the trucks are working 8 hours per day.

Resource and Activity Auto Compute Actuals

When a Resource **Auto Compute Actuals** field is unchecked, the work for a resource may be read from the Primavera Timesheet system or manually entered.

But when the **Activity Auto Compute Actuals** field is checked, this makes all the activity resource assignments to be **Auto Compute Actuals**, irrespective of their settings in the Resource Window.

When the user uses the **Apply Actuals** function and activities or resources are set to **Auto Compute Actuals**, Primavera calculates the Remaining Units based on the Remaining Duration and the Actual Units by subtracting the Remaining Units from Budgeted Units.

There are several places where the Resource **Auto Compute Actuals** field in Primavera is displayed:

❖ Against each resource in the **Resources Window, Details** tab,

❖ In a column in the **Resources Window**, when displayed, and

❖ Against each resource after it has been assigned to an activity and is displayed in the **Resources** tab of the bottom window in the **Activities Window**.

The option may **ONLY** be switched on or off against a resource in the **Resources Window** and if changed in the **Resources Window** it will affect all resource assignments for all projects for this resource.

The **Activity Auto Compute Actuals** field may be displayed as a column in the **Activities Window**:

Activity ID	Activity Name	Auto Compute Actuals	Activity % Comp	Orig Dur	Rem Dur
NEWPROJ (New Project)					
A1000	Project Management	☑	30%	20d	15d

General	Status	Predecessors	Successors	Relationships	Expenses	Codes

Activity	A1000		Project Management	
Resource ID Name		Auto Compute ...	Default Units / Time	Rate Type
PM.Project Manager		☐	8h/d	Price / Unit

Calculate Costs from Units

With this option checked, the costs for a resource are calculated from the **Resource Unit/Time** when a resource is assigned to an activity. When unchecked, the costs remain at zero when a resource is assigned to an activity. This was called **Cost Units Linked** in earlier versions of P6.

When a resource has been assigned to an activity, there is a Resource Assignment field available in the **Resources** tab of the **Activities Window** titled **Calculate Costs from Units**. This is checked to match the **Calculate Costs from Units** field in the **Resources Window**.

The **Activities Window** field titled **Calculate Costs from Units** is not linked to the **Calculate Costs from Units** field in the **Resources Window** and only adopts the setting when a resource is assigned to an Activity.

19.3.6   Units and Prices Tab

Effective Date and Rates

Each Resource may have up to five rates (Price/Unit) and these rates may be varied over time.

❖ The **Effective Date** represents a change in Rate or availability at that point in time.

❖ To display the other rates, their columns should be displayed in the **Units and Prices** tab.

❖ The column titles of **Price/Unit 1** to **Price/Unit 5** may have their descriptions edited in the **Admin, Admin Preferences..., Rate Type** tab. These titles are shared with Roles.

❖ When a rate is added the effective date is the date from which the rate is applied.

Shifts

Resource Shifts are used in conjunction with leveling and resources should not be assigned unless they are being used.

NOTE: It is the author's opinion that shifts are better modeled using calendars and a separate activity for each shift.

Afternoon and night shifts may be made LOE activities and linked to the day shift activity and schedule logic made between only the day shift activities.

The picture below shows a day and night shift example:

NOTE: See the authors article **Scheduling with Shifts in Primavera P6** for more examples.

19.3.7 Roles Tab

❖ A Resource may be assigned more than one Role, and their **Proficiency** for the Role, in this tab.

❖ When multiple Roles are assigned, one is assigned as the **Primary Role**.

Role ID	Role Name	Proficiency	Primary Role
Oz.BM	Bid Manager	3 - Skilled	☑
Oz.SC	Scheduler	4 - Proficient	☐

19.3.8 Notes Tab

Notes may be added here but there are no Notebook Topics available.

19.3.9 Progress Reporter Tab

Originally called **Progress Reporter** tab, then changed to **Timesheets** and now back to **Progress Reporter** and has been removed from P6 Professional.

When Timesheets are implemented, the user must be added as a system user through the **P6 Web Administer**, **User Access** form where the user is assigned to a P6 Resource, thus providing the link from the timesheet user to the Primavera resource. This is covered in the **Admin Menu**, **Users** section.

For timesheets to operate, the **Uses timesheets** box in the **Progress Details** tab must also be checked and the **Timesheet Approval Manager** selected.

19.3.10 Role Rate Vary Over Time – P6 Version 20 Enhancement

Resources have always been able to vary their rate over time but Roles were not and this was one of the drivers to use Resources and not Roles in P6 project schedules.

P6 Version 20 introduced and enhancement allowing Role Rates to vary over time.

When an activity spans two periods then P6 calculates a proportional rate based on the activity duration in each period. You must use the **Recalculate Cost Assignment** function during or after rescheduling.

The pictures below show the result of assigning the Role Foreman above and has 4 days work in 2020 and 6 in 2021 and then 3 and 7 days. They demonstrate how the cost are calculated which are not entirely correct as the unit rate should be lower in 2021 and higher in 2022.

		Dec 28							Jan 04							
Sat	Sun	Mon	Tue	Wed	Thr	Fri	Sat	Sun	Mon	Tue	Wed	Thr	Fri	Sat	Sun	Mon

28-Dec-20 ▬▬▬▬▬▬▬▬▬▬▬▬▬▬▬▬▬ 08-Jan-21

01-Jan-21

		Dec 28							Jan 04							
Sat	Sun	Mon	Tue	Wed	Thr	Fri	Sat	Sun	Mon	Tue	Wed	Thr	Fri	Sat	Sun	Mon
		1	1	1	1	1			1	1	1	1	1			
		$112	$112	$112	$112	$112			$112	$112	$112	$112	$112			

		Dec 28							Jan 04							
Sat	Sun	Mon	Tue	Wed	Thr	Fri	Sat	Sun	Mon	Tue	Wed	Thr	Fri	Sat	Sun	Mon

29-Dec-20 ▬▬▬▬▬▬▬▬▬▬▬▬▬▬▬▬

		Dec 28							Jan 04							
Sat	Sun	Mon	Tue	Wed	Thr	Fri	Sat	Sun	Mon	Tue	Wed	Thr	Fri	Sat	Sun	Mon
			1	1	1	1			1	1	1	1	1			1
			$114	$114	$114	$114			$114	$114	$114	$114	$114			$114

NOTE: This is the same way that resources calculate when an Activity spans an **Effective Date** and the P6 Version 2021 **Cost Spread** enhancement is covered in the next paragraph.

19.3.11 Resource and Role Cost Spread Consider Rate Changes Over Time

The **Admin Preferences, Options** has introduced in Version 21 a **Cost Spread** option.

This changes how the cost spreads for resources and role rates are reflected in the Resource Usage Spreadsheet and Profile, Activity Usage Spreadsheet and Profile, Tracking View, Resource Assignments window, and Activity Usage Spreadsheets and Profiles, Publishing and Reports.

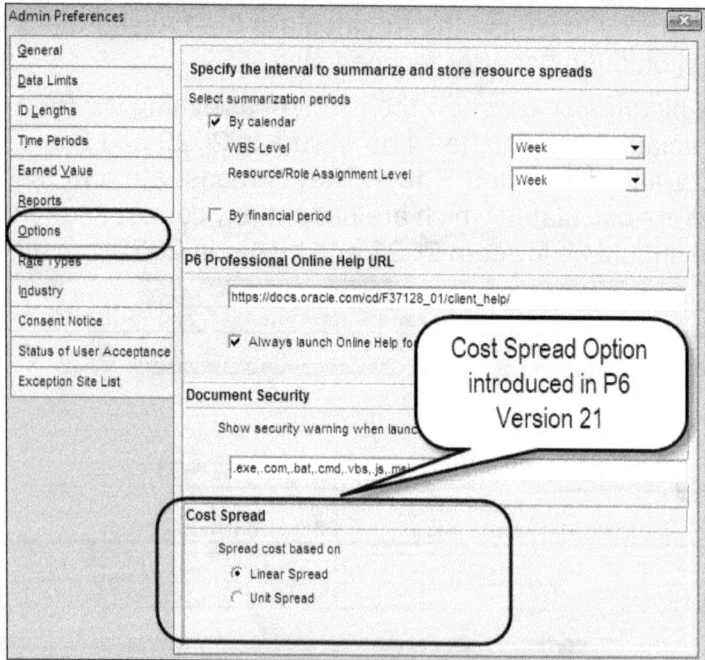

In earlier versions when a resource rate changed during the duration of an activity then P6 took an average rate for the resource over the activity duration which gave an incorrect cashflow.

❖ In the example below I have created a resource that doubled its unit rate from $20.00 per hour to $40.00 per hour after the first week:

Resource ID		Resource Name	Resource Type
CS		Cost Spread	Labor

General | Codes | Details | Units & Prices | Roles | Notes | User Defined Fields

Shift Calendar: [____] ... Shift: 1

Effective Date	Max Units / Time	Standard Rate
01-Jan-22	8/d	$20/h
17-Jan-22	8/d	$40/h

❖ With **Linear Spread** option, which is how older versions calculated, an average rate of $30.00 per hour is used, which is $240.00 per day:

Activity ID	Original Duratio	03				Jan 10							Jan 17				
		hr	Fri	Sat	Sun	Mon	Tue	Wed	Thr	Fri	Sat	Sun	Mon	Tue	Wed	Thr	Fri	Sat
Cost Spreads																	
A1000	10																

Display: Curr... | Display: Acti... | Display: Open Projects Only

Resource ID	Activity ID	At Completion Cost				Jan 10							Jan 17				
			Sun	Mon	Tue	Wed	Thr	Fri	Sat	Sun	Mon	Tue	Wed	Thr	Fri	Sat
CS				$240	$240	$240	$240	$240				$240	$240	$240	$240	$240
	A1000			$240	$240	$240	$240	$240				$240	$240	$240	$240	$240

❖ With **Unit Spread** option, the cash flow is calculated correctly:

Activity ID	Original Duratio	03				Jan 10							Jan 17				
		hr	Fri	Sat	Sun	Mon	Tue	Wed	Thr	Fri	Sat	Sun	Mon	Tue	Wed	Thr	Fri	Sat
Cost Spreads																	
A1000	10																

Display: Curr... | Display: Acti... | Display: Open Projects Only

Resource ID	Activity ID	At Completion Cost				Jan 10							Jan 17				
			Sun	Mon	Tue	Wed	Thr	Fri	Sat	Sun	Mon	Tue	Wed	Thr	Fri	Sat
CS				$160	$160	$160	$160	$160				$320	$320	$320	$320	$320
	A1000			$160	$160	$160	$160	$160				$320	$320	$320	$320	$320

20   ASSIGNING ROLES, RESOURCES AND EXPENSES

20.1   Understanding Roles, Resources and Expenses

During the planning stage, **Roles** may be assigned to Activities to gain an understanding of the long-term resource demand, and they are later replaced by a **Resource** when it is known who will be undertaking the work.

If you are not using named resources then you should consider not using Roles, as Resources have more functionality than Roles. A Resource may be assigned:

❖ Directly to an Activity, or

❖ To a Role which has been assigned to an Activity.

There are three types of resources, **Labor**, **Nonlabor** and **Material**, as discussed in the previous chapter.

❖ A **Labor Resource** has additional functionality including Overtime, Resource Calendars, Shifts and user defined Autocost rules.

❖ The **Labor** and **Nonlabor** resources are similar to the Microsoft Project **Work Resources**.

❖ A **Material** resource is similar to Microsoft Project **Material Resources** but may not have the units displayed in **Activities Window** columns.

Primavera also has a function titled **Expenses**:

❖ Costs may be assigned to activities without resources and may be assigned a quantity and the default quantity is one.

❖ This function is similar to the **Cost Resource** function in Microsoft Project.

❖ As the project progresses, Actual and To Complete Units and Costs may be assigned to Expenses in the same way as resources.

❖ Expenses may have Actual costs assigned before the activity has started,

❖ Expenses may have Remaining Costs when the activity is complete, and

❖ Expenses may be assigned to Milestones.

Expense units may NOT

❖ Have their units displayed in **Activities Window** columns, or

❖ Be assigned **Resource Curves**.

20.2 Understanding Resource Calculations and Terminology

A Resource has three principal components after it has been assigned to an Activity:

❖ **Quantity**, in terms of **Work** in hours or days or **Material** quantities required to complete the activity, which are referred to as **Units** by Primavera,

❖ The **Resource Unit Rate** is termed **Price/Unit** in Primavera, and

❖ **Cost**, which is calculated from the **Resource Unit Rate** x **Units**.

Each Resource and Expense has the same four fields for **Costs** and **Units**, which are **Budget**, **Remaining**, **Actual** and **At Completion**. The relationship among these fields changes depending on whether the activity is **Not Started**, **In progress** or **Complete**.

❖ When an activity is **Not Started** and the % **Complete** is zero then:

  ➢ **Budget** is normally linked to **Remaining** and **At Completion,** therefore a change to one will change the other two and they will always be equal, and

  ➢ **Actual** will be zero.

❖ When the activity is marked Started and would normally be In progress and the % Complete is between 0.1% and 99.9% then:

> **Budget** becomes unlinked from **Remaining** and **At Completion**, thus allowing progress and the **At Completion** value to be compared to the **Budget** value (of the current schedule), or a **Baseline Budget** value or a **Baseline At Completion** value, and

> **At Completion** = **Actual** + **Remaining** and have a link to **Units % Complete**, where a change in value to one will result in a change to the other values.

❖ When the activity is **Complete** and the **Units % Complete** is 100% then:

> **Remaining** is set to zero, and

> **At Completion** = **Actual**.

The Budget values for Costs and Units are linked to the At Completion values until:

❖ An Activity has been marked as Started or has a % Complete, or

❖ The **Link Budget and At Completion for not started activities** in the **Project Window Calculations** tab is unchecked,

Calculations	
Activities	
Default Price / Unit for activities without resource or role Price / Units	$0/h
☐ Activity percent complete based on activity steps	
☑ Link Budget and At Completion for not started activities	

20.3   Project Window Resource Preferences

Preferences set in the **Project Window** decide how each individual activity and resource is calculated. These Preferences and Defaults (which may be changed for each resource assignment) affect how all resources in a project are calculated. They are set in the **Project Window** and pertain to all activities and resources.

20.3.1   Resources Tab

Resources	
Assignment Defaults	
Specify the default Rate Type for new assignments	
Price / Unit ▼	
☑ Drive activity dates by default	
Resource Assignments	
☑ Resources can be assigned to the same activity more than once	

Assignment Defaults

There are five Resource Rates available in Primavera. One rate may be set as a project default. After assignment to an activity, the Resource Rate may be changed using the **Rate Type** field in the **Resources** tab of the **Activities Window**.

Drive activity dates by default

This is covered in more detail in the next section.

Resource Assignments

Checking the **Resources can be assigned to the same activity more than once** box enables a resource to be assigned to an activity more than once. This is useful if it is required to assign a resource at the beginning of an activity and later at the end of an activity with a lag.

For example, one may want to assign a crane on the first day of the activity to assist in erecting, and one the last day to assist in dismantling. This check box needs to be checked for a resource to be assigned twice to an activity.

20.3.2 Understanding Resource Option to Drive Activity Dates

A resource has the following fields that are linked and a change to the **Original lag** or **Original Duration** will make a change to one or both dates:

❖ **Original Lag**. The duration from the **Activity Start Date** to the **Resource Start Date**, which is the date the resource commences work.

❖ **Original Duration**. The duration that a resource is working.

❖ **Start**. The **Resource Start Date** = **Activity Start Date** + the **Resource Original Lag**.

❖ **Finish**. This date is calculated by the addition of the **Activity Start Date** + **Original Lag** + **the Original Duration.**

When the **Drive Activity Dates** option is switched off it is possible for a resource to calculate outside the activity duration. In the following example the activities are 5 days long and the resources assigned to each activity are working for 10 days. This has resulted in the resource being overloaded. The Resources acknowledge the activity Start date but not the Finish Date.

	Jan 31							Feb 07							Feb 14					
S	M	T	W	T	Fri	S	S	M	T	W	T	Fri	S	S	M	T	W	T	Fri	S

	Jan 31							Feb 07							Feb 14					
S	M	T	W	T	Fri	S	S	M	T	W	T	Fri	S	S	M	T	W	T	Fri	S
	8	8	8	8	8			16	16	16	16	16			8	8	8	8	8	
	8	8	8	8	8			8	8	8	8	8								
								8	8	8	8	8			8	8	8	8	8	

20.3.3 Calculations Tab

The **Calculations** tab in the **Projects Window**:

Calculations	
Activities	**Resource Assignments**
Default Price / Unit for activities without resource or role Price / Units $50/h	When updating Actual Units or Cost
	○ Add Actual to Remaining
☐ Activity percent complete based on activity steps	● Subtract Actual from At Completion
☑ Link Budget and At Completion for not started activities	☑ Recalculate Actual Units and Cost when duration % complete changes
○ Reset Original Duration and Units to Remaining	☐ Update units when costs change on resource assignments
● Reset Remaining Duration and Units to Original	☑ Link actual to date and actual this period units and costs

Activities – Default Price/Unit for activities without resource Price/Units.

This rate is also used to calculate the resource costs when an activity is not assigned roles or resources, but is assigned a quantity in the **Activities Window, Status** tab.

NOTE: This works in the same way as Microsoft Project.

The other functions in this tab affect the updating of resourced activities and are covered in the **Updating a Resourced Schedule** chapter.

20.3.4 User Preferences Applicable to Assigning Resources

Units/Time Format

Select the **Time Units** tab. The **Units/Time Format** enables Microsoft Project-style formatting of **Resource/Time Format** showing Resource utilization as a percentage or as units per duration.

Units/Time Format
Resource Units/Time can be shown as a percentage or as units per duration
○ Show as a percentage (50%)
● Show as units/duration (4h/d)

Resource Assignments

The Calculations tab has two Resource Assignment options:

> **Resource Assignments**
>
> When adding or removing multiple resource assignments on activities
>
> ⦿ Preserve the Units, Duration, and Units/Time for existing assignments
>
> ○ Recalculate the Units, Duration, and Units/Time for existing assignments based on the activity Duration Type

Preserve the Units, Duration, and Units/Time for existing assignments. With this option, as Resources are added or deleted the total number of hours assigned to an activity increases or decreases. Each Resource's hours are calculated independently.

Recalculate the Units, Duration, and Units/Time for existing assignments based on the activity Duration Type. The total number of hours assigned to an activity will stay constant as second and subsequent resources are added or removed from an activity.

NOTE: This function does not work when the Duration Type is **Fixed Duration and Units/Time**. It is recommended that **Preserve the Units, Duration, and Units/Time** for existing assignments be used as a default as each individual resource assignment does not change as resources are added or removed from an activity.

20.3.5   Assignment Staffing

The **Assignment Staffing** option is self-explanatory and should be considered carefully when resources and roles have different rates. If it is not understood and set correctly the resource may end up with the incorrect unit rate when assigned to a Role or existing Resource.

When two users have different settings, this may result in a schedule having two different rates for the same resource.

20.4 Activities Window Resource Preferences and Defaults

20.4.1 Details Status Form

This form has a section titled **Labor Units** at the right side as seen in the following picture. The drop-down menu enables you to select which data is to be displayed in this section of the form.

There is a link between the entries in this form and the values that are assigned to resources:

❖ The values in this form are the sum of the values assigned to Resources and Roles.

❖ When these values are edited, they will change the values assigned to Resources and Roles.

NOTE: It is possible to enter a **Labor Unit** value in the **Status** tab and not assign a resource. When a resource is assigned the resource will adopt this value in the **Status** tab. This rate is set in the **Calculations** tab in the **Projects Window, Activities tab – Default Price/Unit for activities without resource or Roles Price/Units** field. When a resource is reassigned then the units will be adopted from the value assigned to the activity and the resource unit rate will be reassigned to the activity.

20.4.2   Activity Type

There are five **Activity Types** assigned in the **General tab** in the **Activities Window**. In summary:

Task Dependent

Activities assigned as Task Dependent acknowledge their Activity Calendar when scheduling and the Finish Date is calculated from the Activity Calendar. Resources ignore their Resource Calendar and are scheduled on the Activity Calendar.

Resource Dependent

Activities assigned as Resource Dependent acknowledge their **Resource Calendar** when being scheduled.

The Activity Finish Date is calculated based on the longest Resource Duration when the resource option of **Drive Activity Dates** is checked against the resource assignment.

NOTE: The activity start date calculated on the activity calendar, not the resource calendar, may delay the start of an activity when the resource calendar has longer working hours than the activity calendar.

Level of Effort (LOE)

This Activity Type spans other Activities, using relationships. Therefore, the Start Date, Finish Date, and Durations may change as the start or finish date of activities that it is dependent upon change during scheduling or updating.

This type of activity does not create a critical path irrespective of the float calculations that are displayed.

Resources assigned to a Level of Effort activity are not considered in calculations when a schedule is **Leveled**.

Level of Effort activities may not be assigned a **Constraint**.

When creating a **LOE** activity and the bar is not displayed, check the **Bars** form to ensure a LOE bar has been created and is being displayed.

Activity Type	Dec 13	Dec 20	Dec 27	Jan 03
	M T W T F S S	M T W T F S S	M T W T F S S	M T W T F
Level of Effort				
Level of Effort				
Level of Effort				
Task Dependent				
Task Dependent				

Start Milestone

This Activity Type is used to indicate the commencement of a Phase, Stage, or a major event in a project.

❖ It has only a Start Date and no Duration or Finish Date.

❖ It may only have Start Constraints assigned.

❖ It may not have time-dependent resources assigned but may have an **Owner** and **Primary Resource** assigned without effort.

Finish Milestone

This Activity Type is used to indicate the completion of a Phase, Stage, or a major event in a project.

❖ It has only a Finish Date and no Duration or Start Date.

❖ It may only have **Finish Constraints** assigned.

❖ It may not have time-dependent resources assigned but may have an Owner and Primary resource assigned without effort.

WBS Summary Activity

This is an activity that spans the duration of all activities which are assigned exactly the same WBS Code and, unlike a Level of Effort Activity, do not have any predecessors or successors.

Activity ID	Activity Name	Activity Type		Qtr 2, 2005			Qtr 3, 200	
			Mar	Apr	May	Jun	Jul	Aug
Hardware								
AS-H	Hardware Summary	WBS Summary						
AS310	Site Preparation	Task Dependent						
AS240	Installation Begins	Start Milestone						
AS315	Install Electrical Pow...	Task Dependent						
AS109	Test & Debug Line A	Task Dependent						
AS110	Test & Debug Line B	Task Dependent						
AS111	Pilot Start Line A	Task Dependent						
AS112	Start-Up Line B	Task Dependent						
AS275	Path Refinement an...	Task Dependent						
AS265	Path Refinement an...	Task Dependent						

Therefore, a WBS activity will change duration when either the earliest start or latest finish of activities that it spans is changed. This may happen as the project progresses and activities do not meet their original scheduled dates, or the duration of an activity is changed, or logic is changed, or the schedule is leveled.

It is similar to the way Summary activity durations are calculated in Microsoft Project, except the activities do not need to be demoted, as in Microsoft Project.

WBS activities may be used for:

❖ Reporting at summary level by filtering on WBS activities,

❖ Entering estimated costs at summary level for producing cash flow tables while the detailed activities are used for calculating the overall duration for the WBS and day-to-day management of the project, and

❖ Recording costs and hours at summary level when is it not desirable or practical to record at activity level, especially when the detailed activities are liable to change.

NOTE: It does not matter how activities are Grouped as they always span activities with the same WBS Code.

20.4.3   Duration Type

Setting the Duration Type

The **Duration Type** becomes effective after a resource has been assigned to an activity.

The **Duration** Type for all new activities is set in the **Defaults** tab in the **Projects Window** and all new activities are assigned this Duration Type.

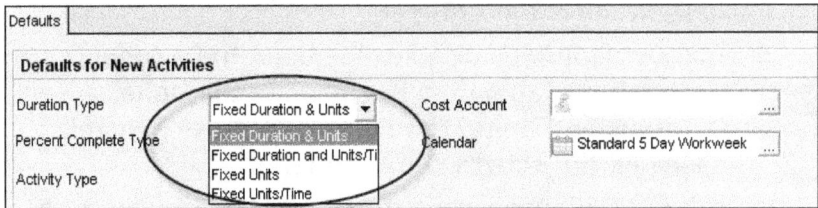

The **Duration Type** for each new activity may be changed in the **General** tab in the **Activities Window** or by displaying the **Duration Type** column:

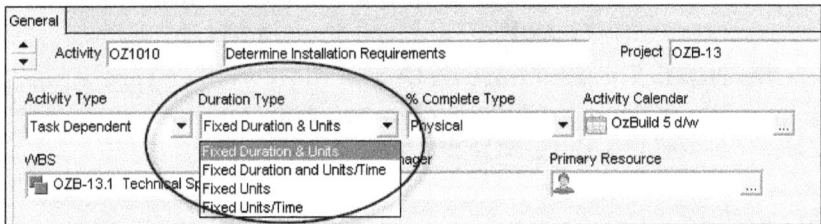

The **Duration Type** determines which of the following variables change when one of the others is changed in the equation:

❖ **Resource Units = Resource Units per Time Period** x **Duration**

When an activity is in progress, this equation is modified to:

❖ **Remaining Resource Units = Resource Units per Time period** x **Remaining Duration**

Fixed Duration & Units

❖ This option is used when the Duration of an activity should not change when Resources are added or removed, or Units/Time changed.

❖ A change to the Duration will change the Units/Time; however, the Units will remain constant.

Fixed Duration & Units/Time

❖ This Duration Type disables the **User Preferences, Calculations tab option Recalculate the Units, Duration, and Units/Time for existing assignments based on the activity Duration Type**.

❖ This option is used when the **Duration** of an activity should not change when Resources are added or removed, or **Units/Time** changed.

❖ A change in the **Duration** will change the **Units**; however, the **Units/Time** will remain constant.

❖ **NOTE:** The **Estimate at Completion** WILL change when the activity duration is changed and the number of resources WILL NOT change.

Fixed Units

❖ This option is used when the amount of work required to finish an activity is constant.

❖ Changing the **Duration** or the **Units/Time** will not change the number of hours required to complete the activity.

Fixed Units/Time

❖ This option is used when the same number of people are required to complete an activity irrespective of the activity duration.

❖ Changing either the **Units** or the **Duration** will not change the **Units/Time**.

Authors Recommendation

The duration of both **Fixed Units** and **Fixed Units/Time** activities will change if the resource Units/Time Period or Remaining Units are changed. It is the author's preference for activity durations not to change when editing resources and recommends:

❖ **Fixed Duration & Units** when the estimate at completion must not change, and

❖ **Fixed Duration & Units/Time** when the crew size must remain constant.

20.5 Assigning and Removing Roles

To assign a Role to an activity:

❖ Select the one or more activity to be assigned the Role,

❖ Select the **Resources** tab in the **Activity Details** form,

❖ Click on the ⬜ **Roles... Assign** toolbar icon to open the **Assign Roles** form,

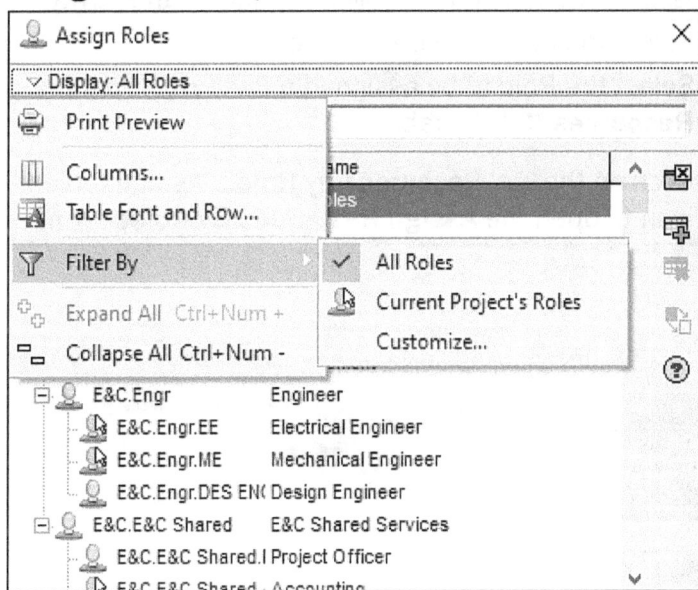

- ❖ Use the **Display:**, **Filter By** menu to select either:
  - ➢ **All Roles**, which will display all Roles in the database, or
  - ➢ **Current Project's Roles**. This option will only display Roles that have been assigned to this project, or
  - ➢ **Customize**, which opens a **Filter** form enabling the user to limit the number of displayed Roles by creating a filter.
- ❖ At this point, the Roles hours and costs may be edited as required.

20.6 Assigning and Removing Resources

Resources may be assigned directly to:

- ❖ An activity that has an Assigned Role, or
- ❖ An Activity without a Role.

20.6.1 Assigning a Resource to an Assigned Role

To assign a Resource to a Role assigned to an activity:

- ❖ Select the activity to be assigned a Resource,
- ❖ Select the Role to be assigned a Resource from the **Resources Details** tab,
- ❖ Click on the 🔲 **Resources by Role... Assign** toolbar icon to open the **Assign Resources By Roles** form,

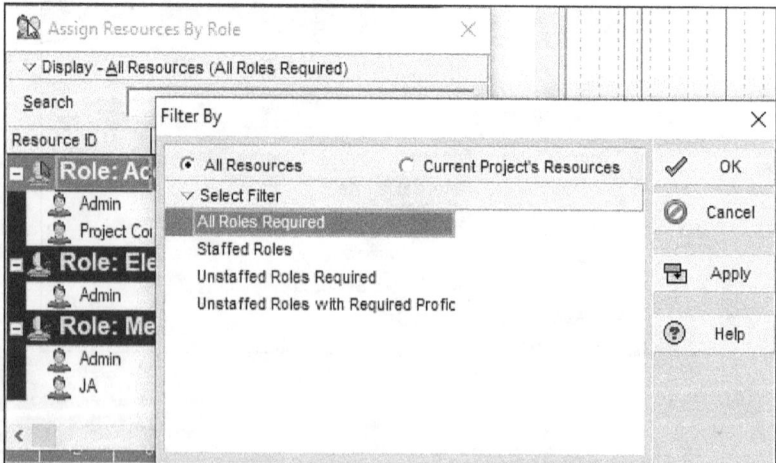

❖ Click on the **Display:** menu and select **Filter By** to open the **Filter By** form,

➤ **All Roles Required**: Chooses to view all roles assigned to the activity.

➤ **Staffed Roles:** Displays Roles with an assigned resource.

➤ **Unstaffed Roles Required:** Displays Roles without an assigned resource.

➤ **Unstaffed Roles with Required Proficiency:** Displays Roles without an assigned resource and requires a resource with a specific proficiency level.

❖ Select which **Resources** you wish to have displayed in the **Assign Roles** form from the **Filter By** form,

❖ Select [🖼 Apply] to return to the **Assign Resources By Role** form,

❖ From the **Assign Resources By Rol**e form click the **Resource** you wish to assign,

❖ To assign the **Resource** either:

➤ Double-click the Resource, or

➤ Click on the [🖼] icon.

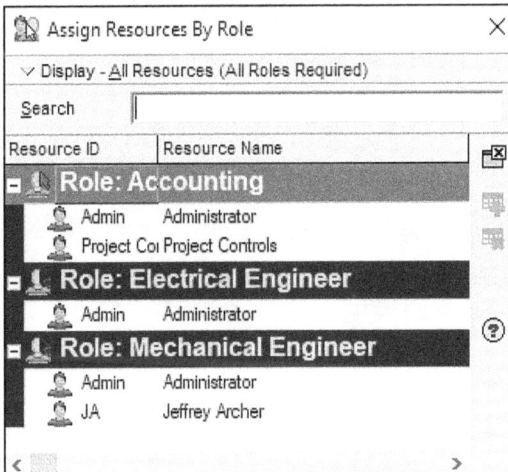

20.6.2 Assigning a Resource to an Activity Without a Role

To assign a Resource to an activity:

❖ Select the activity to be assigned the Resource,

❖ Click on the 🖼 **Resources... Assign** toolbar icon to open the **Assign Resource** form,

❖ Click on the **Display:** menu and select **Filter By** and then select from the three options, which resources you wish to display in the **Assign Resources** form,

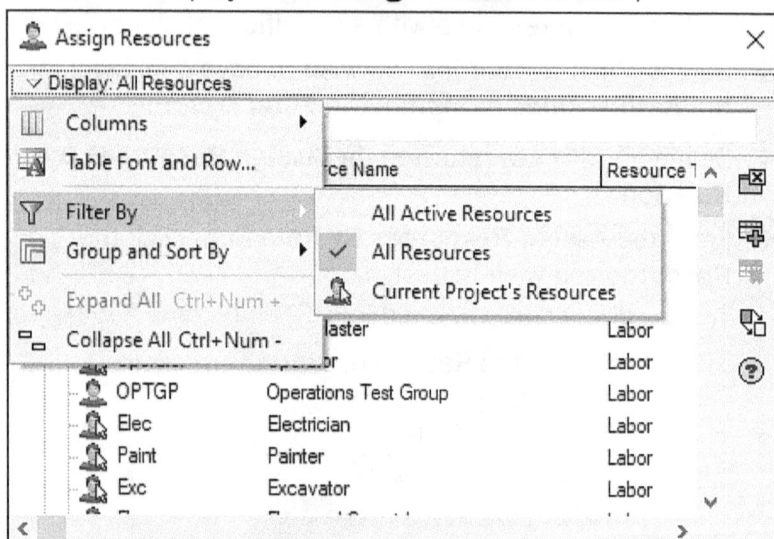

❖ To assign the Resource either:

➢ Double-click the Resource, or

➢ Click on the 🖼 icon.

You may now edit the hours or Units/Time Period for each resource.

20.6.3   Removing a Resource

Before you remove a Resource from an activity that has more than one resource assigned to it, you must be aware of your **Resource Assignment** preferences. These preferences determine if the total number of Units assigned to the activity (or work) will be reduced or remain constant as resources are deleted.

To remove a resource, select one or more Resources in the bottom pane Resource tab and either:

❖ Strike the **Del** key, or

❖ Click on the ⊞ Remove icon at the bottom of the screen, not the ☒ on the **Edit** toolbar.

After the last resource is removed there will be the message:

Confirmation

All labor resources have been deleted on this activity. Do you want to reset labor units to zero on this activity?   ✔ Yes   ✗ No

☐ Do not ask me about this again

❖ If you select ✔ Yes then the **Resource Units** values in the **Activities Window**, **Status tab** will be set to zero, and the Resource Costs in the **Activities Window, Status tab** will be calculated from the value entered.

If you select ✗ No then:

❖ The Units values in the **Activities Window, Status** tab will be set to equal the Resource values before they were deleted, and

❖ The Cost values in the **Activities Window, Status** tab will be calculated from the value set in **Calculations** tab in the **Projects Window**.

When you assign a resource to an activity in this condition the resource will adopt the Units value from the **Activities Window**, **Status** tab, ignoring the **Default Units /Time Period** set in the **Resource Window,** but normally calculate the resource value from the resource Rate.

❖ At this point you will have Units and Costs assigned to an activity that may be seen in the **Activities Window**, **Status tab** without any assigned resources, which may not be desirable.

∨ Labor Units		∨ Labor Cost	
Budgeted	80h	Budgeted	$4,000
Actual	0h	Actual	$0
Remaining	80h	Remaining	$4,000
At Complete	80h	At Complete	$4,000

NOTE: When you assign a resource to an activity in this condition the resource will adopt the Units value from the **Activities Window**, **Status tab**, ignoring the **Default Units /Time Period** set in the **Resource Window,** but will normally calculate the resource value from the resource Rate.

20.6.4 Assigning a Resource to an Activity More Than Once

The option in the **Projects Window Resources** tab under the **Resources Assignments** heading enables a resource to be assigned more than once to an activity.

Resources

Resource Assignments

☑ Resources can be assigned to the same activity more than once

For example, a resource could be assigned to work at the start of an activity and then in conjunction with **Resource Lag** work again at the end of an activity.

20.7 Resource and Activity Duration Calculation and Resource Lags

20.7.1 Activity Duration

An Activity Duration (or Activity Remaining Duration of an In progress Activity) is adopted from the longest Resource Duration (or Resource Remaining Duration of an In progress Activity) when more than one resource has been assigned to an activity.

In a situation where more than one Resource has been assigned to an activity with different Units and/or Units/Time, the Resources may have different durations.

In the following example the Activity Duration is 10 days, which is calculated from David William's **Resource Original Duration** of 10 days:

Role ID	Resource ID Name	Original Lag	Original Duration	Remaining Units	Remaining Units / Time
Oz.SE	ARL.Angela Lowe	0d	5d	40.00h	100%
Oz.BM	DTW.David Williams	0d	10d	40.00h	50%
Oz.CS	MAY.Melinda Young	0d	5d	40.00h	100%

20.7.2 Resource Lag

A Resource may be assigned a Lag, the duration from the start of the activity to the point at which the Resource commences work.

In the following example the Activity Duration is 12 days, which is calculated from Angela Lowe's **Resource Original Lag** of 7 days and **Resource Original Duration** of 5 days:

20.8 Expenses

Expenses are intended to be used for one off non-resource types such as travel costs or the engagement of a consultant.

Expenses may be created using the:

❖ **Expenses Window** and assigned to an activity, or

❖ Created in the **Expenses tab** of an activity.

20.8.1 Expenses Window

The **Expenses Window** is opened by:

❖ Clicking in the ▦ icon on the Project toolbar, or

❖ Selecting Project, Expenses.

Creating a new **Expense** is similar to creating a new activity:

❖ Select **Edit**, **Add**, and

❖ The **Select Activity** form will then be displayed and the activity the expense is to be associated with is selected.

Expense Item	Expense Category	Vendor
Training Manuals	Training	Eastwood Harris Pty Ltd
Primavera Training Course	Training	Eastwood Harris Pty Ltd

General | Activity | Costs | Description

Expense Item	Expense Category
Training Manuals	Training

Vendor
Eastwood Harris Pty Ltd

Cost Account	Document Number
Con.11.4 Training	110803 P6V81S

Enter the required information in the tabs in the bottom window.

20.8.2 Expenses Tab in the Activities Window

This tab may have all the columns of data available in the **Expenses Window** displayed. All the fields may be edited from this tab:

Activity ID	Activity Name		January 2014				February 2014				March 2014			
			06	13	20	27	03	10	17	24	03	10	17	24
P6 Training														
A1000	Primavera Training Course													

Expenses

Activity A1000   Primavera Training Course   Project Training

Expense Item	Accrual Type	Price / Unit	Unit of Measure	Remaining Units	Budgeted Cost
Primavera Training Course	Uniform over Activity	$1,800	day	3.000	$4,800
Training Manuals	Uniform over Activity	$100	each	10.000	$1,000

20.9 Suggested Setup for Creating a Resourced Schedule

The simplest calculation options should be used as a default, and more complex options considered only when there is a specific scheduling requirement.

The following lists the processes and suggested options that could be considered when creating a resourced schedule. It is important to set all the parameters before the activities are added, otherwise a lot of time is wasted changing parameters on a number of activities. These are not intended to suit every project but are a starting point for less experienced users.

- ❖ Set the **Units/Time** format by selecting **Edit, User Preferences...** to open the **User Preferences** form and select the **Time Units** tab.
  - ➢ There is a choice of percentage (50%) or units/duration (4h/d). This should be set on personal preference.
  - ➢ The author prefers (4h/d) as this reduces typing.
- ❖ Set the **Resource Assignments** option by selecting **Edit, User Preferences...** to open the **User Preferences** form and select the **Calculations** tab.
  - ➢ It is suggested that the **Preserve the Units, Duration, and Units/Time for existing assignments** is selected. With this option, as Resources are added or deleted, the total number of hours assigned to an Activity increases or decreases. Each Resource's hours are calculated independently.
  - ➢ The options under **Assignment Staffing** need to be carefully considered and understood so that when Resources are assigned to Roles and resource assignments are changed, that the user understands which Unit Rate and which Unit Cost will remain against the activity.
- ❖ In the **Project Window, Default**s tab set the default **Activity Type**.
  - ➢ It is suggested that **Task Dependent** is used, as with this option, Resource calendars are not used, making the schedule simpler.
- ❖ In the **Project Window, Defaults** tab set the default **Duration Type**.
  - ➢ It is suggested that **Fixed Duration & Units** is used. With this option the Activity Duration does not change when resource assignments are altered, and when an Activity Duration is changed, the Units do not change, so your estimate of hours and costs will not change.

❖ In the **Project Window, Default**s tab set the default **Percent Complete Type**.

 ➢ The author prefers to use **Physical** as this enables the Activity Percent Complete to be independent of the Activity Durations.

❖ In the **Project Window, Resource** tab set the default **Resource Assignment Defaults**.

 ➢ Unless multiple Rates are being used, then **Price/Unit** should be selected.

 ➢ Check **Drive activity dates by default**.

21 RESOURCE OPTIMIZATION

The schedule may now have to be resource optimized to:

❖ Reduce peaks and smooth the resource requirements, thus reducing the mobilization and demobilization costs, or to reduce the demand for site facilities, or

❖ Reduce resource demand to the available number of resources, or

❖ Reduce demand to an available cash flow when a project is financed on income.

21.1 Reviewing Resource Loading

There are a number of facilities for reviewing resource loading, which consist of either displaying an **Activity Layout**, or **Resource Assignments Window** or the **Tracking Window** or running a report.

21.1.1 Activity Usage Spreadsheet

This window is displayed by clicking on the 🖳 icon or selecting **View**, **Show on Bottom, Activity Usage Spreadsheet**.

❖ This displays a total of all the resource costs or units assigned to activities:

❖ Right-clicking will display a menu and the **Spreadsheet Fields...** option allows the selection of **Cumulative** and **Time Interval** display of **Resource** and **Expenses** information.

© *Eastwood Harris*

NOTE: Cumulative **Expense Units** and **Material Resources Units** are not available in this view, only **Costs**.

Spreadsheet Options... allows the calculation of the average number of resources:

The units are formatted using the **User Preferences**, **Time Units** tab. If the minimum time unit is an hour, ensure the **User Preferences**, Resource Analysis **Interval for time-distributed resource** calculations is set to one hour; otherwise, the data will not be displayed correctly when the timescale is opened up to hours:

21.1.2   Activity Usage Profile

This is displayed by clicking on the 📊 icon or selecting **View**, **Show on Bottom, Activity Usage Profile.**

❖ It displays the total resource histogram for selected or all activities. Right-click the Histogram for the options:

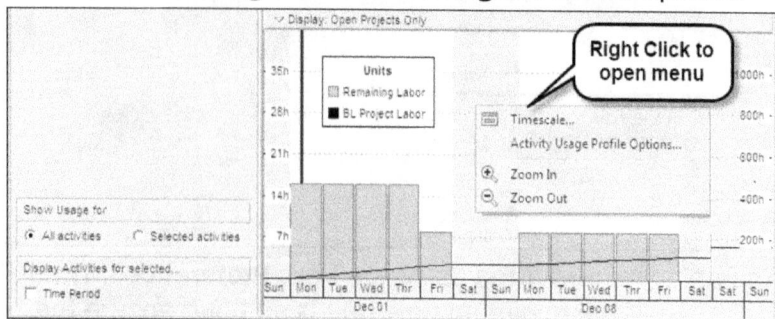

❖ The Activity Usage Profile Options... menu opens up the Activity Usage Profile Options form:

21.1.3   Resource Usage Spreadsheet

This is displayed by clicking on the [icon] icon or selecting **View**, **Show on Bottom, Resource Usage Spreadsheet**.

❖ This form has three windows showing the resources that are assigned to activities.

❖ Each window has a menu when right-clicking in the window.

❖ The units are formatted using the **User Preferences**, **Time Units** tab.

❖ As with the **Activity Usage Spreadsheet**, when the minimum time unit is an hour, ensure that the **User Preferences**, **Resource Analysis Interval for time-distributed resource calculations is set to one hour**; otherwise, the data will not be displayed correctly.

❖ When multiple resources are selected on the left-hand window then the corresponding Resource activities are displayed in the center and right-hand side window:

Version 15.2 introduced saving the selected Resources in both the **Resource Spreadsheet** and **Resource Histogram** when you save a Layout.

21.1.4 Editing the Resource Usage Spreadsheet – Bucket Planning

This new option in Primavera Version 6.0 enables resource assignment values to be manually edited. This enables more control over the assignment of resources that are working intermittently on an activity.

This is similar to editing a Microsoft Project Resource Usage table and making a resource assignment "Contoured."

The following picture shows the edited values in the **Resource Usage Spreadsheet**.

Activity ID	R U Dec	Jan	Feb	Mar	Apr	May	Jun	Jul	Aug	Sep	Oct
∨ Display: Activity Resource ...	∨ Display: Open Projects Only						2012				
Innovative Constr	176	176	3	1	6		7		8	200	18
Design Engineeri	176	176	3	1	6		7		8	200	18
Structural Engineers	176	176	3	1	6		7		8	200	18
A1000	176	176	3	1	6		7		8	200	18

Each time period, therefore, may contain a different value.

NOTE: It is recommended that you experiment with this function if you plan to progress **Bucket Planned Resources** as the author has found this process gives some interesting results for the incomplete portion of an in progress activity.

21.1.5 Resource Usage Profile displaying a Resource Histogram

Click on the ▨ icon or select **View**, **Show on Bottom**, **Resource Usage Profile**.

❖ The options in this form are similar to the ones covered in the previous paragraphs,

❖ Stacked or individual histograms are available from the menu:

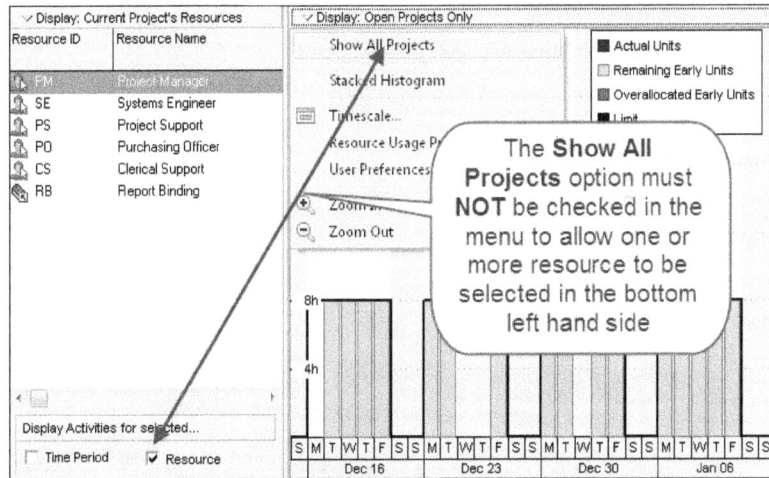

The **Show All Projects** option must NOT be checked in the menu to allow one or more resource to be selected in the bottom left hand side

Version 15.2 introduced saving the selected Resources in both the **Resource Spreadsheet** and **Resource Histogram** when you save a Layout.

NOTE: The resource availability is displayed using the **Resource Calendar** when the **Activity Type** is set to **Task Dependent** and the activity is scheduled using the Activity Calendar. At this point in time there may be overload indicated when the Activity Calendar has more working time than the Resource calendar.

21.1.6 Histogram Bars Exact Values – P6 Version 20 Enhancement

In P6 Version 20 the value of histograms may be displayed on histograms in the **Resource Usage Profile**, **Activity Usage Profile**, and **Tracking View** by selecting the **Show Values** option in the **Profile Option form**, **Graph** tab:

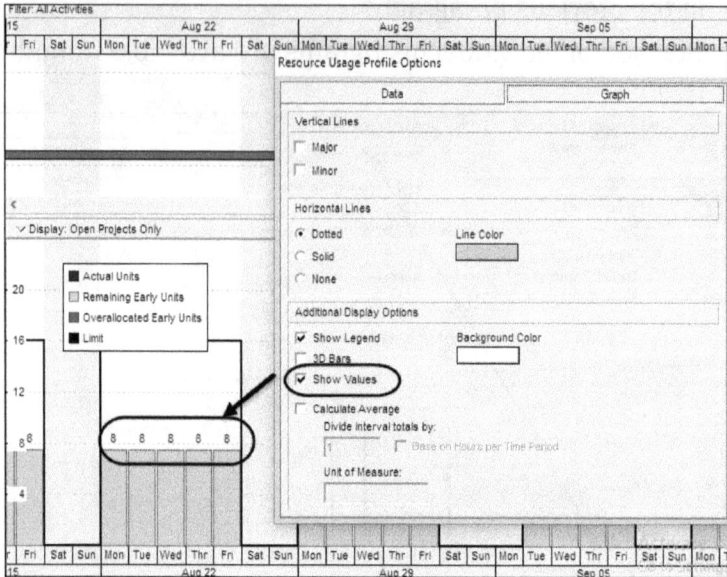

21.1.7   Stacked Histogram

Stacked Histograms may be created, but the process is a little long winded:

❖ Display the **Resource Usage Profile**,

❖ Filter on the **Current Project resources**,

❖ Right click in the Histogram area, which will be blank, select **Stacked Histogram**,

❖ Right click and select **Resource Profile Options** and

❖ Add the **Resources** as displayed in the picture below:

21.1.8 Activity Usage Profile Displaying S-Curves

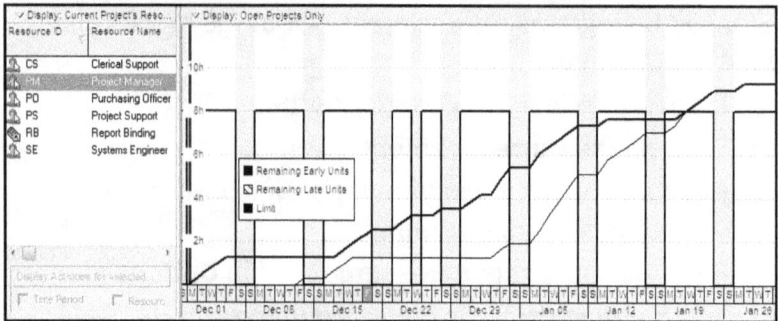

You must be prepared to experiment with the formatting menus by right-clicking in each of the windows of the above displays to understand all the many options, which include:

❖ Roles or Resources,

❖ All Resources, All Active Resources and Current Projects Resources only,

❖ Options to show period and cumulative values, or an average by dividing by a number,

❖ Options to filter, and

❖ Options to Group and Sort.

NOTE: The formatting is a little disappointing as:

❖ The bars and curves share the same color and are difficult to see when both are displayed,

❖ There is no control over the vertical scale, so it is hard to compare one resource profile with another and

❖ Only one profile ay be displayed.

21.2 Resource Assignments Window

The **Resource Assignments Window** has some functions that are very useful, especially when you wish to copy and paste data into Excel.

This view is essentially a time-phased view that is grouped by default by Resource, Role, or Activity and allows the display of:

❖ Cumulative and Period totals

❖ Cost of all Resource Types

❖ Units of all Resource Types

NOTE: This view does not show either **Expense Costs** or **Expense Units**. So, using this view for a cash flowing project with Expenses will not give the full value of the project. Resource Units Totals, say at project level, are only available when one resource type is displayed by using a filter.

21.3 Copying and Pasting into Excel

The following data may be copied and pasted into Excel:

❖ Activity data from the Activities Window
❖ Activity Usage Spreadsheet
❖ Tables in the Tracking Window
❖ Resource Assignments Window.

The **Resource Usage Spreadsheet** may NOT be copied and pasted into Excel, but similar data may be obtained in the **Resource Assignments Window** and copied and pasted.

You should be aware of the following issues:

❖ The **User** Preferences need to be appropriate, especially for date formatting, if you wish the data to be pasted as dates into Excel.

❖ Dates that are pasted with an "A" at the end may be removed with the Excel command of Find and Replace. You will need to put a space before the "A", so you do not lose the "A" in front of August.

❖ To remove the "*" at the end of a date you must use the syntax of "~*" in the Find and Replace command as a "*" on its own will replace all the data in the spreadsheet.

21.4 Other Tools for Histograms and Tables

Oracle Primavera also sells a reporting add-on software package titled **Primavera Earned Value Management** which allows the production of a number of reports, such as time-phased table, bubble, and period variance.

Contact your local Oracle Primavera distributor or go to the Oracle Primavera web site for more information.

There are many non Oracle software package listed on **www.primavera.com.au** that you may wish to investigate.

21.5 Methods of Resolving Resource Peaks and Conflicts

Methods of resolving resource overload problems are:

❖ **Revising the Project Plan.** Revise a project plan to mitigate resource conflicts, such as changing the order of work, contracting work out, or using off-site pre-fabrication, etc.

❖ **Duration Change.** Increase the activity duration to decrease the resource requirements, so a 5-day activity with 10 people could be extended to a 10-day activity with 5 people. If the activity is **Fixed Duration and Units** this will calculate as required.

❖ **Resource Substitution.** Substitute one resource with another available resource.

❖ **Increase Working Time.** This may release the resource for other activities earlier and is created by working more days per week or hours per day.

❖ **Split an activity around peaks in demand.** Some software enables the activity splitting, which enables work to be split around peaks in resource demand. This function is not available in Primavera, however, an activity may be split into two individual activities to allow the work to cease in times of peak demand. If one needs to relate back to a baseline then two new activities may be created to represent the split and the original activity made into a hammock to span the two new activities but remember to display the LOE Baseline Bar the baseline is set as **All Activities** in the **Bars** form:

Display	Name	Timescale	Filter	Preview
☑	Primary Baseline	Primary Baseline Bar	All Activities	
☐	Primary Baseline	Primary Baseline Bar	Milestone	▽ ▽
☐	Primary Baseline	Primary Baseline Bar	Summary	

❖ **Leveling the schedule.** This technique delays activities until resource(s)are available.

❖ **Resource Curves** or **Manually Editing the Resource Spreadsheet** may assist in some instances.

21.6 Resource Leveling

21.6.1 Methods of Resource Leveling

After resource overloads or inefficiencies have been identified with Resource and Tables, the schedule may now have to be leveled to reduce peaks in resource demand. Leveling is defined as delaying activities until resources become available. There are several methods of delaying activities to level a schedule:

❖ **Turning off Automatic Calculation and Dragging Activities**. This option does not maintain a critical path and reverts to the original schedule when recalculated. This option should not be used when a contract requires a critical path schedule to be maintained, as the schedule may no longer calculate a critical path.

❖ **Constraining Activities**. A constraint may be applied to delay an activity until the date that the resource becomes available from a higher priority activity. This is not a recommended method because the delay of the higher priority activity may unlevel the schedule.

❖ **Sequencing Logic**. Relationships may be applied to activities sharing the same resource(s) in the order of their priority. In this process, a resource-driven critical path is generated. If the first activity in a chain is delayed, then the chain of activities will be delayed. But the schedule will not become unleveled and the critical path will be maintained. In this situation, a successor activity may be able to take place earlier and the logic will have to be manually edited.

❖ **Leveling Function**. The software Resource Leveling function levels resources by delaying activities without the need for Constraints or Logic and finds the optimum order for the activities, based on user defined parameters. Again, as this option does not maintain a critical path developed by durations and relationships, it should not be used when a contract requires a critical path schedule developed in this way. The Leveling function may be used to establish an optimum scheduling sequence and then Sequencing Logic applied to hold the leveled dates and to create a critical path.

The Resource Leveling function enables the optimization of resource use by delaying activities until resources become available, thus reducing the peaks in resource requirements. This feature may extend the length of a project.

The leveling function should only be used by novices with extreme caution.

❖ It requires the scheduler to have a solid understanding of how the software resourcing functions calculate.

❖ Leveling increases the complexity of a schedule and requires a different approach to building a schedule. In principle, the sequencing logic is supplemented by activity Priorities, but a Closed Network should still be maintained.

Your ability to understand how the software operates is important for you to be able to utilize the leveling function with confidence on larger schedules. It is recommended that you practice with small, simple schedules to gain experience in leveling and develop an understanding of the leveling issues, before attempting a complex schedule.

21.6.2 Resource Leveling Function

This section outlines the software Resource Leveling functions, including:

❖ **Level Resources** form,

❖ Guidelines on Leveling, and

❖ What to look for if resources are not leveling properly.

21.6.3 Level Resources Form

The **Level Resources** form enables you to assign most of the Leveling prerequisites. Select **Tools**, **Level Resources...** to open the **Level Resources** form:

Option added in Version 20

❖ **Automatically level resources when scheduling** – levels the schedule each time the schedule is recalculated. **NOTE:** This option is not recommended.

❖ **Consider assignments in other projects with priority equal/higher than**. – levels resources and at the same time considers the demands of other projects. The leveling priority is set in the **Projects** window, **General** tab. 1 is the highest and 100 the lowest priority.

❖ **Preserve scheduled early and late dates** – in simple terms, when unchecked enables the option of Late Leveling. This is explained in more detail in the following paragraphs as the computations are a little more complicated. Late Leveling pushes forward in time activities from their late dates to meet the resource availability and provides the latest dates the activities may be started and finished without delaying the finish date of the project.

❖ **Recalculate assignment costs after leveling** – is used with the resource **Effective date** and **Price/Unit**. These facilities allow a change in the cost of a resource over time. The Resource Costs are recalculated based on the resource **Price/Unit** if an activity is moved into a different price bracket when this check box is marked.

❖ **Level all resources** – if checked, the schedule levels all the resources; if unchecked, enables the **Select Resources** form to be opened and one or more resources selected for leveling.

❖ **Level resources only within activity Total Float**

➢ When checked, the leveling process will not generate negative float but may not completely level a schedule. Thus, the activities will only be delayed until all float is consumed and leveling will not extend the finish date of the project. This option will also check the **Preserve scheduled early and late dates** option.

➢ When unchecked, leveling will allow activities to extend beyond a **Project Must Finish By date**, when assigned in the **Projects Window Dates** tab, or beyond the latest date calculated by the schedule and may create **Negative Float**.

➢ **Preserve minimum float when leveling** – works with **Level resources only within Total Float** and will not level activities if their float will drop below the assigned value.

> **Max percent to over-allocate resources** – works with **Level resources only within activity Total Float** and enables the doubling of the resource availability, although this new limit is not displayed in the histogram limits.

❖ **Leveling priorities** – sets leveling priorities, and activities are assigned resources according to the data item chosen in the first line. If two activities have the same value in the first line, then the priority in the second line is used. The Activity ID is the final value used to assign resources. There are many options for leveling priority and the following are some to consider:

> **Activity Leveling Priority** is a field that may be set from 1 Top to 5 Lowest; the default is 3 Normal. Those with a priority 1 Top are assigned resources first.

> **Activity Codes** or **User Defined Fields** and many other data fields such as **Remaining Duration**, **Early Start**, **Total Float**, and **Late Start** may be used to set the priority for leveling.

21.7 Guidelines for Leveling

Leveling a schedule is a skill that is acquired through practice and experience and there are a few fundamentals that a user must bear in mind before attempting to level a complex schedule.

❖ If you are not an experienced scheduling software user, then it is strongly suggested that you obtain some serious experience in using Primavera with resources before attempting to use leveling on a complex schedule, especially if you are trying to level a progressed schedule. You will need this experience to resolve some of the complex issues that are often present when leveling a schedule.

❖ You need to approach the structure of the schedule differently at the beginning of schedule construction. Without leveling, schedulers normally apply soft logic (sequencing logic) to prevent a number of activities occurring at the same time. If leveling is your method of scheduling, then soft logic should be omitted from the beginning of the construction of the schedule.

❖ All users and reviewers of the schedule must understand that a leveled schedule may dramatically change with the addition or removal or change to activities or change in priorities.

❖ There are some principles that should be considered when leveling:

➢ Only level resources that are overloaded and that you are unable to supplement easily, or that have an absolute limit.

➢ Try leveling one resource at a time and view the histograms to ensure each resource is leveling. If a resource is not leveling and the histograms display overload, you will need to go through the check list on the next page and level again. This process often finds a driving overload resource and leveling that resource levels the whole project.

➢ After all resources are leveling individually, you should start leveling with two resources and then three. Do not start leveling with all the resources at once, as the schedule will often do some drastic things and extend the project end date unrealistically.

➢ Do not expect a perfect result; be satisfied with an average resource usage that meets your requirements over periods, such as months. Sort out small peaks in future resource requirements nearer to the start of the activity.

To understand how leveling will delay or change durations of activities you will need to be aware of which of the above combinations you have employed in your schedule, and you will then need to understand how each combination calculates under a non-leveling environment.

21.8 What to look for if Resources are Not Leveling

It is very frustrating if you have a project that will not level. Try some of these options when your schedule will not level:

❖ Have you selected a resource to level in the **Select Resources** form? The resources to be leveled must be selected in the **Select Resources** form.

❖ Have you set the **Limits** in the **Resource Window**? A resource needs a limit to level.

❖ A resource will not be leveled when you assign a resource to an activity with a Units per time period greater than value set in the resource dictionary. This may occur when:

➤ The **Resource Limit** in the **Resource** window is reduced, or

➤ An activity has been assigned a resource with a **Unit per Time Period** that is greater than the **Limit**, or

➤ When the activity has **Fixed Units** and the duration of an activity has been reduced, thus increasing the assigned **Units per Time Period** over the maximum available in the **Resource** form.

❖ Have you assigned a **Mandatory Constraint** to an unleveled activity? Activities with a Mandatory constraint will not be leveled.

❖ Have you assigned resources to a **WBS** or **LOE** activity? These will not be leveled.

❖ Have you checked **Level resources only within activity Total Float** option? This option enables activities without float to level.

21.9   Resource Curves

Resource Curves enable a non-linear assignment of resources to schedules in the same way as P3 and Microsoft Project. These are often used on long activities where there is not a requirement for a linear assignment of resources.

Resource curves are assigned in the **Curve** column in the **Resources** tab of the **Activities Window**:

The Electrical wiring activity in the following picture has a bell-shaped Resource Curve assigned to it:

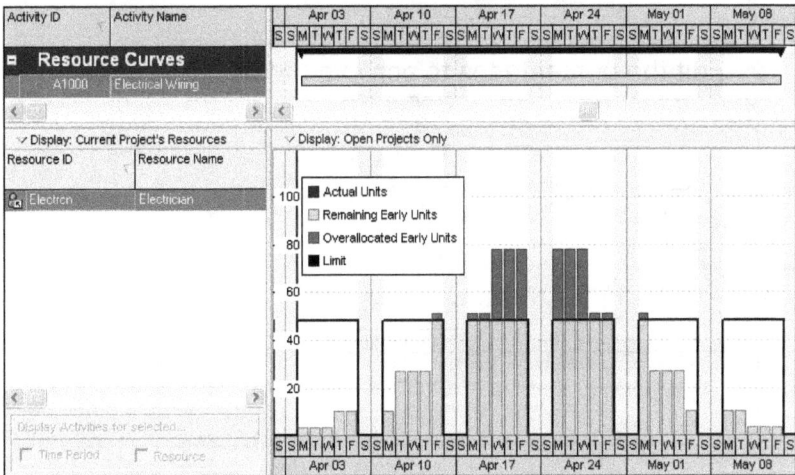

To create and use **Resource Curves**:

❖ Select **Enterprise**, **Resource Curves**... to open the **Resource Curves** form:

❖ Default curves may not be deleted or edited but may be copied in the **Modify Resource Curves** form.

❖ **Global** curves may be edited, copied or deleted.

❖ To create a new curve, select ⊕ Add to open the **Select Resource Curve To Copy From** form and select a curve to copy.

❖ You will be returned to the **Resource Curves** form where the title may be edited.

❖ Click the [Modify...] icon to open the **Modify Resource Curves** form:

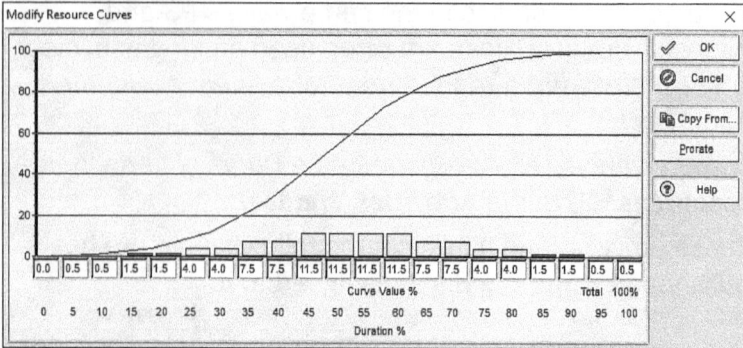

❖ Edit the percentages to achieve the desired shape:

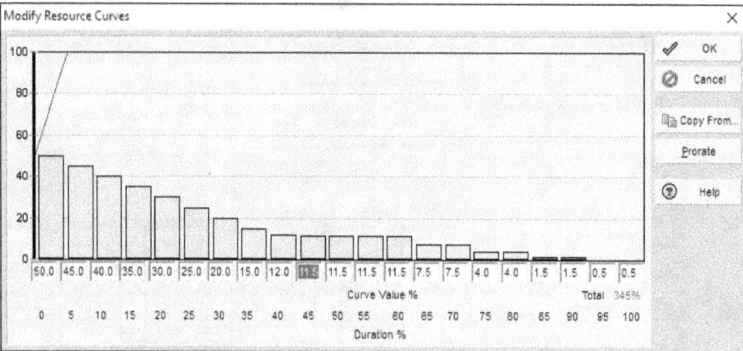

❖ Click on [Prorate] to make the percentages add to 100%:

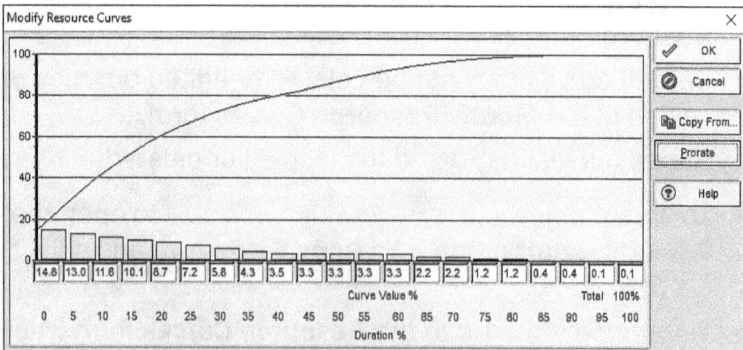

You may now assign this curve to an activity.

22 UPDATING A RESOURCED SCHEDULE

It is often considered best practice to update a project between 10 and 20 times in its lifecycle. Some companies update schedules to correspond with accounting periods, which are normally every month. This frequency is often too long for projects that are less than a year in duration, as too much change may happen in one month. Therefore, more frequent updating may identify problems earlier.

Updating a project with resources employs a number of preferences and options, which are very interactive and will require a significant amount of practice by a user to understand and master them.

After reading this chapter and before working on a live project, inexperienced users should gain confidence with the software by:

❖ Creating a new project and setting the **Defaults**, **Preferences**, and **Options** to reflect the method in which you wish to enter information and how you want Primavera to calculate the project data.

❖ Creating two or three activities and then assigning two or three resources to each activity.

❖ Updating the Activities and Resources as if you were updating a schedule and observe the results.

❖ Alter the preferences and defaults if you are not receiving the result you require. Re-update and note the preferences and defaults for future reference.

Some of these settings may have been set by your organization and you may not be assigned access rights to change the settings. You should still go through the updating process in a test project with dummy data similar to your real project data and be prepared to change those settings to which you do have access, as required.

Updating a project with resources takes place in two distinct steps:

❖ The dates, durations and relationships are updated using the methods outlined in the **Updating an Unresourced Schedule** chapter, and

❖ The Resource, Expenses Units (hours and quantities) and Costs, both the Actual to Date and To Complete, are then updated. The Resource values may be automatically updated by Primavera from the % Complete or imported from accounting and timesheet systems or updated by the Primavera Timesheet system.

A decision needs to be made about what data is to be entered or imported into the schedule and what data is to be calculated by the software and the software options set appropriately.

22.1 Understanding Budget Values and Baseline Projects

22.1.1 Cost and Units Budget Values

The Budget Values in Primavera are assigned to both Units and Costs for each Resource and Expense at the time the Resource or Expense is assigned to an Activity.

Budget Values reside in the current project and in all Baseline Projects.

The Budget values by default are linked to the At Completion values when an activity has not commenced but after the activity is in progress by being marked as Started or having a % Complete these values become unlinked. Budget Values, like Planned Dates, may contain irrelevant data, depending on how you update your project schedule.

NOTE: The author recommends always ignoring the Budget Values in both the Current Project and any Baseline, and users should always compare the At Completion Costs and Units of the Current Project with the At Completion Costs and Units of a Baseline project.

22.1.2   Baseline Project and Values

A Baseline project is a complete copy of a project including the relationships, resource assignments and expenses.

The Baseline values are values against which project progress is measured. All these values may be read by and compared with the current project values and show variances from the original plan.

A Baseline would normally be created prior to updating a project for the first time.

The Primavera Variance columns use Baseline data from Baseline Projects to calculate variances.

22.2   Understanding the Current Data Date

The **Data Date** is a standard scheduling term. It is also known as the **Review Date, Status Date, Report Date, As of Date, Time Now**, and **Update Date**.

❖ The **Data Date** is the date that divides the past from the future in the schedule. It is not normally in the future but is often in the recent past due to the time it may take to collect the information required to update the schedule.

❖ **Actual Costs** and **Quantities/Hours** or **Actual Work** occur before the **Data Date**.

❖ **Costs** and **Quantities/Hours to Complete** or **Work to Complete** are scheduled after the **Data Date**.

❖ **Actual Duration** is calculated from the **Actual Start** to the **Current Data Date** or **Suspend Date**.

❖ **Remaining Duration** is the duration required to complete an activity. It is calculated forward from the **Current Data Date** or **Resume Date** and the **Early Finish** date or an in progress activity is calculated from the **Current Data Date** or **Resume Date** using the:

➢ **Activity Calendar** when the Activity Type is Task Dependent or is Resource Dependent but no Resources have been assigned, or

---

© **Eastwood Harris**

> ➤ **Resource Calendar** when the Activity Type is Resource Dependent and uses the longest Resource Duration.

22.3 Information Required to Update a Resourced Schedule

A project schedule is usually updated at the end of a period, such as each day, week, or month. One purpose of updating a schedule is to establish differences between the plan, which is usually saved as a Baseline, and the current schedule.

The following information is required to update a resourced schedule:

Activities completed in the update period:

❖ **Actual Start** date of the activity,

❖ **Actual Finish** date of the activity,

❖ **Actual Costs** and **Quantities** (Units) consumed or spent on **Labor Resources, Material Resources** and **Expenses.** These may be calculated by the software or collected and entered into the software.

Activities commenced in the update period:

❖ **Actual Start** date of the activity,

❖ **Remaining Duration** or **Expected Finish** date,

❖ **Actual Costs** and/or **Actual Quantities.** These may be calculated by the software or collected and entered into the software.

❖ **Quantities to Complete** and **Costs to Complete.** These may be calculated by the software or collected and entered into the software.

❖ **% Complete.**

Activities Not Commenced:

❖ Changes in Logic, Constraints, or Duration, or

❖ Changes in estimated **Costs**, **Hours** or **Quantities** and

❖ Add or remove activities to represent scope changes.

The schedule may be updated after this information is collected.

Other Considerations

Primavera normally, by default, calculates:

❖ The Units to Complete and in turn the Actual Units, by the relationship between the Remaining Duration and Resource Units.

❖ The Costs to Complete and the Actual Costs by the relationship between the Resource Unit Rate and Resource Units.

When these relationships are turned off, then the Units and Costs may be entered manually.

There are many methods of collecting data to update a project schedule. Traditionally a copy of the schedule was printed out and marked up, but other electronic methods, such as the Primavera Timesheet system or an e-mail based system with spreadsheet or pdf attachments, may be employed to collect the data. Irrespective of the method used, the same data needs to be collected.

It is recommended that only one person update each schedule. There is a high probability for errors when more than one person updates a schedule.

There are many defaults that are discussed in my other P6 books, which we will not go into detail in this book.

NOTE: Essentially, it is recommended not to change any Actual Cost and Unit data when you want the software to calculate the Actual and remaining Cost and Units, as the software normally operates well without any changes.

22.4 Activities Window – Percent Complete Types

There are three % **Complete** types which may be assigned to each activity. The default is adopted from the setting in the **Defaults** tab in the **Projects Window**.

❖ **Physical**

❖ **Duration**

❖ **Units**

As discussed, several times in this book, the author recommends using a default of **Physical**, and then when actual units and costs information is entered, the Activity Percent Complete will not change.

Also, if you are anticipating using **Steps**, then **Physical Percent Complete** must be used.

22.5 Using Steps to Calculate Activity Percent Complete

An activity percent complete may be defined by using steps. A Step is a measurable or identifiable task required to complete an activity.

Steps are useful to update activities that have many components, where the order of completion is not important but the measurement of progress is. Examples of the use of Steps:

❖ Driving of piles, with the Step Weight of each pile being the length of the pile,

❖ Pouring of footings, with the Step Weight being the m^3 of concrete for each footing,

❖ Pulling of electrical cable, with the Step weight being the weight or length of each cable.

In summary, to use steps:

❖ A **Step Template** is optional and should be created for repeatable Steps by selecting **Enterprise, Activity Step Template...** to open the **Activity Step Templates** form.

❖ Add as many steps as required and assign their weight, which will be used to apportion the percent complete of an activity.

❖ Check the **Activity percent complete based on activity steps** check box in the **Projects Window, Calculations** tab,

❖ Select the **Physical** in the % **Complete Type** for each activity that is to be measured by steps, in the **General** tab of **Activities Window**,

❖ Select the **Steps** tab in the **Activities Window**,

❖ Format the columns you wish to display,

❖ Add the number of steps you require, or import from a **Step Template**,

❖ Delete any imported **Steps** from the **Step Template** that you do not need,

❖ Edit the descriptions as required,

❖ Edit the **Step Weight** so the **Step Weight Percent** reflects the desired value of the Step,

❖ Check the **Completed** check box as each step is completed and this will update the percent complete.

Step Name	% Complete	Step Weight	Step Weight Percent	Completed
Specify Document Composition	100%	10.0	10.0	☑
Document First Draft	100%	40.0	40.0	☑
Final Draft and Internal Approval	0%	25.0	25.0	☐
Client Approval	0%	25.0	25.0	☐

❖ The **Remaining Duration** may be updated from the **Step** % **Complete** via the **Physical % Complete** using a **Global Change**.

22.6 Updating the Schedule

22.6.1 Recommended Preferences, Defaults and Options for Updating a Project

Most Primavera options and defaults are good, but there are some that should be changed. The options to be considered and checked before updating a schedule are:

❖ **% Complete Type**

> ➤ It is the author's preference to use **Physical % Complete**. This allows the % of deliverables complete to be measured independently of the resource(s) doing the work, thus allowing a comparison of the deliverables completed against the resources consumed.

> ➤ This must be used for **Steps** to operate.

❖ **Activity Type**

> ➤ Activities with known durations should be set as **Task Dependent** and will use the Activity calendar (not the Resource Calendar) for calculating the finish date of the activity.

> ➤ **Resource Dependent** activities should only be used if there are resource availability issues which may only be resolved by the use of **Resource Calendars**.

> ➤ **Level of Effort** and **WBS** activities are useful but should be avoided by the novice user as these add an additional level of complexity that is not required.

❖ **Project Window Calculations tab**

> ➤ The **Calculations** tab in the **Projects Window** sets some important resource defaults that should be reviewed, understood, and set so the schedule calculates the desired way.

> ➤ If a project is to allow the software to update the resources actual costs and units based on the remaining duration, then these should not be changed.

➢ When Steps are to be used, then the check box **Activity percent complete based on activity steps** must be checked.

➢ Should you wish to update Actual and Remaining resource Costs and Units, then you must uncheck **Recalculate Actual Units and Cost when duration % complete changes.**

➢ The **Link actual to date and actual this period units and Costs option found in the Calculations tab of the Project Window** should be checked if it is intended to **Store Period Performance.**

❖ **Duration Type**

➢ It is the author's preference to use **Fixed Duration and Units**, because the estimate to complete is not altered by changing the **Activity Duration** or **Units/Time.**

➢ On the other hand, if you wish the crew size to remain the same when you change the duration, then you should select **Fixed Duration and Units/Time.**

❖ **User Preference Calculation Tab**

➢ Sets the **Resource Assignments** option in the **User Preferences, Calculation** tab to **Recalculate the Units, Duration, and Units/Time for existing assignments based on the activity Duration Type.** Thus, adding and removing resources will not change existing resource assignments

❖ **Timesheets**

➢ Timesheets may be used to update actuals for none, some, or all resources. Organizations using timesheets should have procedures managing their use. Timesheets are out of the scope of this publication, but if they are being used the actual values should be carefully checked before being applied, to ensure they are logical.

❖ **Resources Cost Calculation**

➤ Resource Costs may be calculated from the Resource Unit Rates for each individual resource assignment.

➤ Each resource assignment has a field titled **Calculate cost from units**. When this is checked the resource costs are calculated from the resource units.

➤ The **Calculate costs from units** check box in the **Resource Window, Details** tab sets the default value for **Calculate cost from units** for new resource assignments.

➤ The two fields are not linked and the resource assignment setting may be changed at any time.

❖ **Resource Window Details Tab**

➤ **Auto Compute Actuals**
This field is linked to all resources assignments. When this option is checked for a resource, Primavera calculates the Remaining Units based on the Remaining Duration and the Actual Units by subtracting the Remaining Units from Budgeted Units.
An unchecked resource assignment option may be overridden by applying the **Activity Auto Compute Actuals** option.

➤ **Calculate Costs from Units**
There is a field available when a resource is assigned to an activity titled **Calculate cost from units.** With this option checked the costs for a resource are calculated from the **Resource Unit/Time** when a resource is added to an activity and whenever the Resource Units are changed.

❖ **General Schedule Options**

➢ One of the more important options to review is the **When scheduling progressed activities use** options, as these affect how out-of-sequence progress is handled. These options should be reviewed to ensure that when the schedule is recalculated you will understand what is happening.

➢ The author prefers **Retained Logic** as this gives a more conservative schedule and those relationships that need editing may be edited to reflect retained logic as required.

❖ **Steps**

➢ Should it be decided to use Steps to update a schedule the **Projects Window Calculations** tab should have the **Activity percent complete based on activity steps** option checked, and the Activity must be assigned **Physical % Complete** in the **General** tab of the Activities Window for each activity.

❖ **Earned value calculation**

➢ The **Admin, Admin Preferences...**, **Earned Value** tab, **MUST NOT BE SET TO** "**Budgeted values with planned dates**" because a Baseline has progress, otherwise the Planned Dates will be displayed in the Baseline and these may contain irrelevant data when the schedule has progress.

➢ The author recommends using **At Completion values with current dates**. Then you are comparing the At Completion of the current schedule with the At Completion Values of the Baseline schedule.

22.6.2 Updating Dates and Percentage Complete

The schedule should be first updated as outlined in the **Updating an Un-resourced Schedule** chapter. In summary, this is completed by entering:

❖ The **Actual Start** and **Actual Finish** dates of **Complete** activities.

❖ The **Actual Start**, **% Complete** and/or **Remaining Duration** of In progress activities.

❖ Adjust **Logic**, **Constraints** and **Durations** of **Un-started** activities.

Before updating the **% Complete**, the **% Complete Type** should be checked to ensure that the Actual and Remaining Durations, Costs, and Units calculate as required. This ideally should be done by setting the project defaults at the time the project is created, and adjusting the settings as activities are added and resources assigned.

22.7 Updating Resources

There are many permutations available for calculating resource data. Due to the number of options available, it is not feasible to document all the combinations available for resource calculation. Resource units and costs may be updated using one of the following methods:

❖ Entering Progress Automatically from the timesheets, in a process titled **Apply Actuals**, or

❖ Using the function titles **Update Progress**. This is **NOT** recommended due to the risk that your **Actual Start** and **Early Finish** may be changed by P6 to the **Planned Dates** when the schedule has progress, or

❖ Entering the data using the **Resource** tab in the **Activities Window**, or

❖ Entering the data using the right section of the **Status** tab in the **Activities Window**.

❖ Importing from Excel. Actual dates and Remaining Durations may be imported but Suspend and Resume may not.

22.8   Updating Expenses

Expenses are updated in a similar way to resources in the **Activities Window**, **Expense** tab. Expenses will not be covered in detail, but here are some notes about Expenses that you may find useful:

❖ Expenses do not automatically update from any % Complete and have to be manually updated.

❖ **NOTE:** The Expense **Auto Compute Actuals** option works only with the **Apply Actuals** function, which is used when bringing in data from the Primavera Timesheets module

❖ Expenses may have a cost assigned before their activity is marked started or complete; resources may not. This is useful to represent contractor's mobilization costs. These are scheduled on the Data Date.

❖ Expenses may have a cost to complete after the activity is marked started; resources may not. This is useful to represent contractor's retention. These are scheduled on the Planned Dates.

❖ Expenses may have a negative cost to complete after the activity is marked complete; resources may not. This is useful to represent contractor back charges. These are scheduled on the Planned Dates:

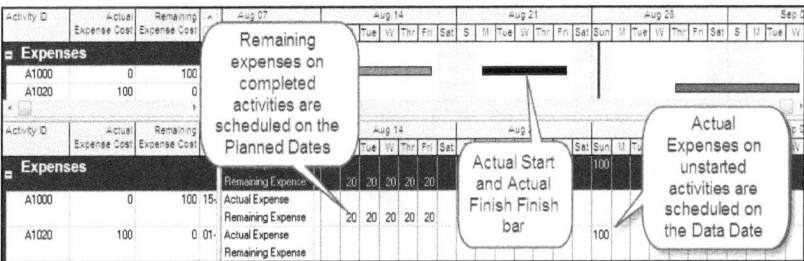

❖ Expenses must be assigned a quantity and unit rate. The quantity is by default a value of one.

❖ Expense quantities may not be displayed in the:

- ➢ **Activities Window** columns, or
- ➢ **Resource Usage Spreadsheet**, or
- ➢ **Resource Usage Profile**, or
- ➢ **Activity Usage Spreadsheet**, or
- ➢ **Tracking Window**, or
- ➢ **Resource Assignment Window**.

❖ Expense Quantities may be displayed in:

- ➢ **Reports,** or
- ➢ **Activity Details**, **Expenses** tab, or
- ➢ **Expenses Window**.

NOTES:

Thus, it is simple to get Expense data into the system but difficult to get **Expense Quantity** data out of the system.

Remaining Expenses assigned to a completed activity are Cash Flowed on the Planned dates, which would normally result in an incorrect Cash Flow, with Remaining Costs in the past.

23 OTHER METHODS OF ORGANIZING PROJECT DATA

The **Work Breakdown Structure – WBS** function was discussed earlier as the main method of organizing projects and activities under hierarchical structures.

There are alternative features available in Primavera for grouping, sorting and filtering activities, resources, and project information including:

❖ Activity Codes

❖ User Defined Fields

❖ WBS Category or Project Phase

❖ Resource Codes

❖ Cost Accounts

❖ Owner Activity Attribute

❖ Assignment Codes – P6 Version 20 Enhancement

❖ Role Codes – P6 Version 20 Enhancement

23.1 Activity Codes

Activity Codes may be used to Group, Sort, and Filter activities from one or more open projects.

❖ **Activity Codes**, such as Phases, Trades, or Disciplines, are often defined in the **Activity Codes Definition** form.

❖ **Activity Code Values** are defined in the **Activity Codes** form, such as:

> ➢ Phases of Design, Procure, Install and Test,

> ➢ Trades of Brickwork, Plumbing and Electrical, and

> ➢ Disciplines of Concrete, Mechanical, Pipework.

❖ **Activity Codes** are assigned from the **Activities Window** using the **Codes** tab in the lower pane or displaying the appropriate **Activity Code** column.

23.1.1 Understanding Activity Codes

There are three types of Activity Codes:

❖ **Global Activity Codes** that may be created at any time and applied to any project.

❖ **EPS** which are created for projects associated with one EPS Node and may only be assigned to project activities that are associated with that EPS Node. Thus, you may wish to create Railway EPS Activity Codes for projects in the Railway EPS and Software Development EPS Activity Codes for projects in the Software Development EPS.

❖ **Project Activity Codes** that may only be created when a project is opened and applied only to the project they were created for. These may be made Global by clicking the ⟨ Make Global ⟩ icon in the **Activity Codes Definition – Project** form.

❖ **NOTE:** Unlike Elecosoft (Asta) Powerproject, only one code per Activity Code dictionary may be assigned to an activity.

23.1.2 Activity Code Creation

This process creates a field in the database where the Activity Codes may be added.

❖ Open an **Activity Codes Definition** form from the **Project, Activity Codes** form by selecting either:

➢ **Global**,

➢ **EPS,** or

➢ **Project**,

Each form is slightly different.

❖ Click the ⟨ Modify... ⟩ icon to open the **Activity Codes Definition** form.

❖ The Activity Codes may be created, deleted, or made into Global and reordered in these forms.

23.1.3 Defining Activity Code Values and Descriptions

Defining an Activity Code is similar to creating a Code in Elecosoft (Asta) Powerproject or renaming a Microsoft Project Custom Outline Code:

❖ From the **Activity Codes** form select **Global**, **EPS** or **Project**,

❖ Select the **Activity Code** to edit from the drop-down box,

❖ Add **Activity Codes Values** and **Descriptions** in the same way as **WBS Codes** and descriptions.

❖ The **Activity Code Color** may be used with the **Timescaled Logic Diagram** in Version 8.1 and 8.2 or **Visualizer** in Version 8.3 and later versions, but not in the Windows PPM client.

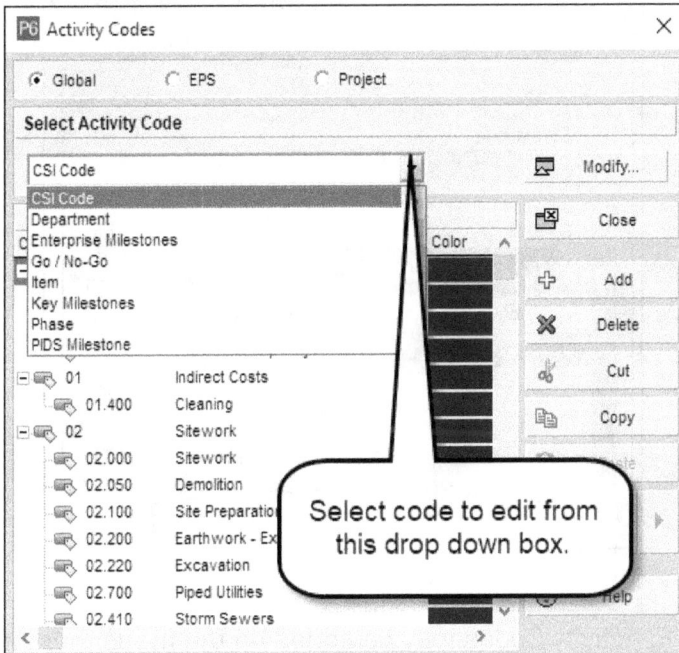

23.1.4   Assigning Activity Code Values to Activities

Activity Codes may be assigned to an activity:

❖ Select the **Codes** tab in the lower pane by clicking the

   [🔲 Assign] icon to open the **Assign Activity Codes** form
   and assign an Activity Code, or

❖ Display the appropriate activity code column and either:

   ➢ Type in the code, or

   ➢ Click twice on the Activity Code cell and open the
      Select "Code" form.

NOTE: Activity Codes may be added on the fly, as there is a new icon titled **New** on the **Assign Activity Codes** form that allows Activity Codes to be created as they are assigned.

23.1.5 Grouping, Sorting and Filtering with Activity Codes

When more than one project is open, an Activity Code may be used to group activities from all the open projects under one code structure.

Activity Codes are Grouped and Filtered in the same way as WBS codes.

23.1.6 Importing Activity Codes with Excel

If an Activity Code is to be imported with activities using the Primavera Excel Import function, the Code must exist in the database before it is imported; otherwise, the code will not be imported.

23.2 User Defined Fields

User Defined Fields may be used for recording information about the data field, as an alternative to Activity Codes and other predefined Primavera fields. The type of data that may be assigned to User Defined Fields would be equipment number, order number, variation or scope number, road, railway or pipeline chainages; address and additional cost data.

NOTE: User Defined Fields are database fields and are available to all projects and importing projects with UDFs will pollute your database with a lot of unwanted data.

Activity data may be filtered, grouped, and sorted using these User Defined Fields in a similar way to Activity Codes.

Data may be imported into the fields and, unlike Activity Codes, the data item does not have to exist in the database before importing.

There are a number of predefined fields that may be renamed and new ones may be created. User Defined Fields may be defined for:

➢ Activities

➢ Activity Resource Assignments

➢ Activity Steps

- ➤ Issues
- ➤ Project Expenses
- ➤ Projects
- ➤ Resources
- ➤ Risks
- ➤ WBS
- ➤ Work Products and Documents

The fields are assigned a **Data Type** from the following list:

- ➤ Text – maximum of 255 characters
- ➤ Start Date and Finish Date – which may be used to create User Definable Field bars
- ➤ Cost
- ➤ Indicator – select from ⊗ ⚠ ⊘ ★
- ➤ Integer
- ➤ Number

After some data has been entered against a field in any project, the **Data Type** may not be changed.

In P6 Version 19 and earlier **User Defined Fields** had to be displayed in column for them to be populated. P6 Version 20 introduced a **User Defined Field Details** tab is now available in many workspaces such as Projects, Resources, Activities and WBS. This is extremely useful as it allows users to see and assign UDFs without having to add extra columns to their Layouts:

One advantage of **User Defined Fields** over **Notebook Topics** is that they may be also displayed in columns and be cut and pasted into other programs like Excel.

Also, **User Definable Field** data may easily be imported from Excel and will not change your project data. You may consider importing data into User Defined Fields and then Global Change the information into the appropriate location as a second step.

Thus, Resource data needs to be imported into Resource User Defined Fields, and Activity data needs to be imported into Activity User Defined Fields.

NOTE: You must be careful that you do not make a User Definable Field with the same field name as a P6 field, otherwise you will not know which is which when creating filters.

Consider placing a full stop after the User Definable field name.

23.3   WBS Category or Project Phase

The **WBS Categories** is assigned to **WBS Nodes** in the **WBS Window** and may be used to Group and Sort WBS Nodes under a different set of headings.

This would enable, for example, all design WBS Nodes that were distributed throughout a project WBS to be grouped together under one heading without assigning an Activity Code to each activity.

Separate set of project WBS Codes can be listed under a dummy code as the project name:

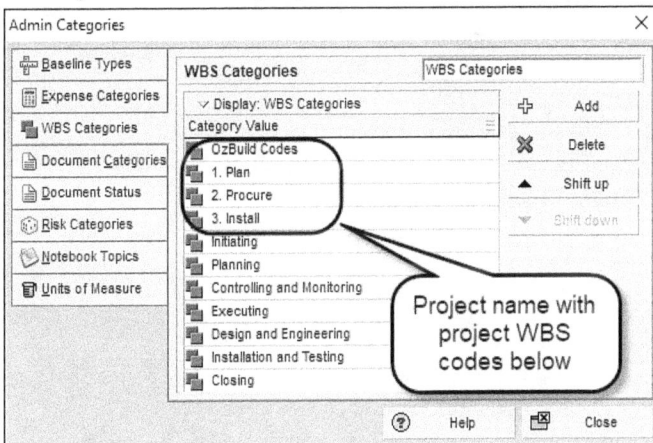

NOTE: WBS Codes are to a WBS in same way as Activity Codes are to Activities and Project Codes to Project but there is only one dictionary or set of WBS Codes.

23.4   Resource Codes

Resource Codes are to resources as Activity Codes are to activities and allow resources to be Grouped, Sorted, and Filtered by these codes. Resources may have codes such as Office, Location, or Employment Status assigned to them.

To create a Resource Code:

❖ Select **Enterprise, Resource Codes...** to open the **Resource Codes** form.

❖ The Resource Codes are created, edited, and deleted in a similar way to Activity Codes.

Resource Codes may be Assigned to Resources in a similar way to Activity Codes by:

❖ Opening the Resources Window,

❖ Displaying the appropriate Code Column, or

❖ Opening the **Codes** tab in the **Resources Window**.

23.5   Cost Accounts

Cost Accounts are to resource assignments as Activity Codes are to activities and are intended to reflect the accounting code structure of a project.

They enable the grouping and reporting of resource data into Cost Accounts which would allow budgets to be calculated and used to update Corporate Budgets.

Cost Accounts have additional functions that Activity Codes do not have:

❖ A default Cost Account for each new Resource or Expense may be specified in the Projects Window, Defaults tab. This is used for each new Resource or Expense and does not affect existing assignments. The Project Default Cost Account may be changed at any time:

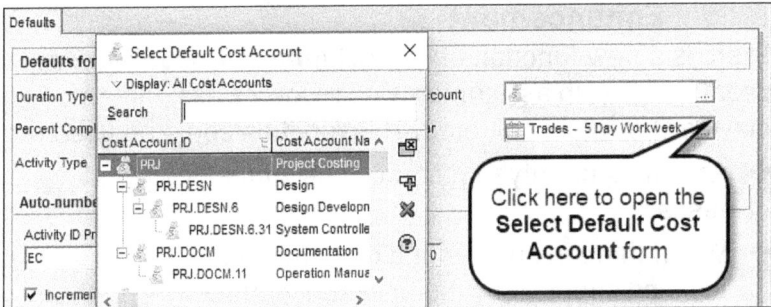

Click here to open the **Select Default Cost Account** form

❖ Cost Accounts may be reassigned and merged.

❖ Cost Accounts may have descriptive fields when they are created.

Costs accounts are created:

❖ In the Professional version **Cost Accounts** form by selecting **Enterprise, Cost Accounts...** and opening the **Cost Accounts** form, and

❖ In the Optional Client by selecting **Administer, Enterprise Data, Activities, Cost Accounts**.

Cost Accounts are assigned to Resources or Expenses by displaying the Cost Account column in the **Activities Window** lower pane **Resources** and **Expenses** tabs.

NOTE: The problem with Cost Accounts is that every resource in a project is assigned the same Cost Account which is not normally how Cost Accounts operate. Usually resources are assigned a cost account such as permanent material or temporary material or hires equipment for accounting style reporting and depreciation. Thus, it makes more sense to use a Resource Code as a Cost Account if more than one Cost Account is to be used in a project.

23.6 Owner Activity Attribute

"Owner," the new activity field in Primavera Version 6.0, enables a user who is not a resource to be assigned to an activity. This now enables the person responsible for an activity to be assigned from the list of users. This function may be used in combination with a Reflection project.

23.7 Assignment Codes – P6 Version 20 Enhancement

There is a new function titled **Assignment Codes** allowing users to code up assignments so resource assignments may be Grouped or Filtered. This function could be used for:

❖ Assigning priority to resource assignments against tasks,

❖ When there is one generic resource in the database that is being supplied by multiple subcontractors and then an **Assignment Code** could be used to identify the subcontractor,

❖ When there is one generic resource in the database and you require a specific skill or qualification for the resource then this could be identified with an **Assignment Code**.

Assignment Codes are created in the same way as other codes by selecting **Enterprise, Assignment Codes**:

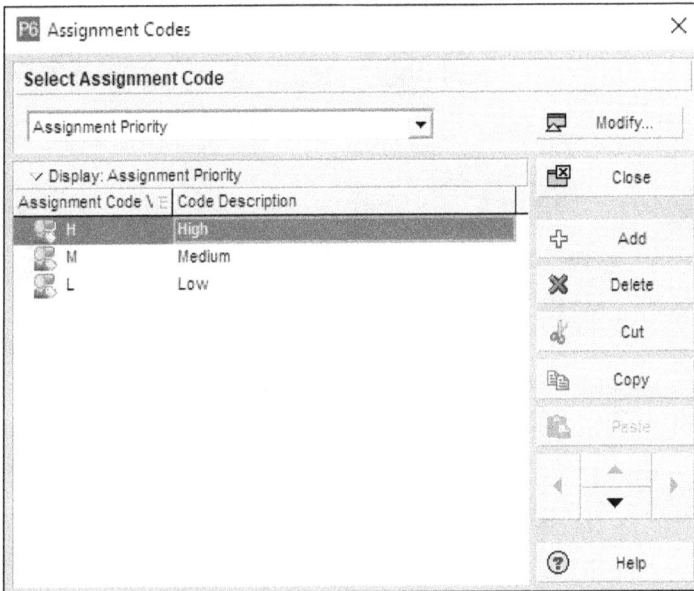

Assignment Codes are assigned to Resource or Role Assignments in the:

❖ **Codes** tab in the **Resource Assignment** window, or

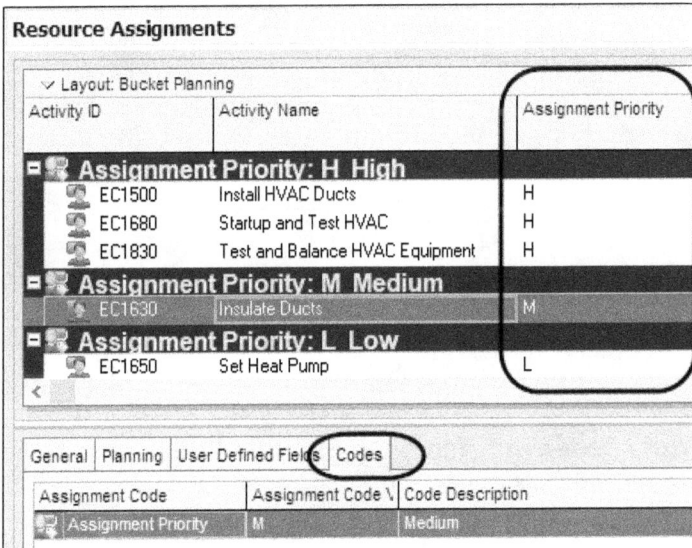

❖ Displaying the **Assignment Code** column in the **Activities** window **Details** pane, **Resources** tab, or

Role	Resource ID Name	Assignment Priority	Budgeted Units
Civil/Structural Crews	Elec.Electrician	M	50
Civil/Structural Crews	HVAC.HVAC	M	50
Civil/Structural Crews	Operator.Operator	M	17

Activity EC1670 — Relocate HVAC Chiller

❖ Displaying the **Assignment Code** column in center section of the **Resource Usage Spreadsheet**.

23.8 Role Codes – Version 20 Enhancement

P6 Version 20 introduced **Role Codes** allowing Roles to be assigned a Code in the same way as Resources and Activities.

This would be useful to assign attributes such as location or office and may be used to group or filter roles.

❖ **Role Codes** are created in **the Role Codes** form by selecting **Enterprise**, Role Codes,

❖ **Role Codes** are assigned to **Roles** in the **Roles** form.

24 GLOBAL CHANGE

24.1 Introducing Global Change

Global Change is a facility for changing more than one data item in one step. Examples of uses of Global Change are:

❖ Assigning or replacing Resources to Roles,

❖ Increasing or decreasing durations of selected activities by a factor,

❖ Creating new activity descriptions or Activity IDs by placing activity codes at the beginning or end of the original value,

❖ Removing constraints,

❖ Changing Calendars.

After you understand the basics you will then develop some interesting ways of using Global Change.

24.2 The Basic Concepts of Global Change

A Global Change may be created, saved, and used at a later date.

A Global Change may not be "Undone."

Select **Tools**, **Global Change...** to open the **Global Change** form:

Name	Available To	User
Assign Resources to Roles	All Users	
Increase Cost by 10% for Resource	All Users	
Increase Durations	All Users	
Top Down - Decrease Units	All Users	
Top Down - Increase Units	All Users	

Buttons: Close, Apply Change, New..., Delete, Modify..., Copy, Paste, Import, Export, Help

The **Global Change** form displays the list of Global Changes available in the project.

❖ [Apply Change] enables the effects of a Global Change in the **Global Change Report** before finalizing changes to the project data by selecting [✔ Commit Changes] in the **Global Change Report.**

❖ [➕ New...] creates a new Global Change.

❖ [🖵 Modify...] enables you to modify the highlighted Global Change.

❖ [✖ Delete] deletes the highlighted Global Change.

❖ [📋 Copy] and [📋 Paste] create a copy of an existing Global Change that may then be edited.

❖ [⬇ Import] and [⬆ Export] are used to import from or export to a Global Change from another database in the **Primavera Change File *.PCF** file format.

After creating a Global Change using the [➕ New...] option or [📋 Copy] and [📋 Paste] or by selecting [🖵 Modify...], you will be presented with the **Modify Global Change** form. This is where you select the data to be changed and where the operation to the data is specified.

There are boxes at the top of the form:

❖ **Select Subject Area** enables the option of Activities, Activity Resource Assignments, or Project Expenses, and

❖ **Global Change Name** is the name displayed in the **Global Change** form.

The form has three lower sections. You will need to click into each area and then use [➕ Add] and [✖ Delete] icons to add or remove criteria or operation lines:

❖ **If** area is where you create the criteria for selecting the data on which to operate. This is similar to creating a filter.

❖ **Then** area is where you specify the operation to be applied to the selected data.

❖ **Else** area is where you have an option to specify an operation to data that has not been selected.

❖ [✓ OK] accepts edits to the Change, but does not execute it.

❖ [⊘ Cancel] cancels any edits to the Change.

❖ [🔁 Change] enables you to see the results of your action in a **Global Change Report** before changing the database.

❖ The other commands are self-explanatory and are used to create and edit lines in the Global Change, but you will need to click into the **If** or **Then** or **Else** sections that you wish to work on.

You will find many examples of Global Changes in my other P6 books.

24.3 Important Points

Here are some important points that you must bear in mind when using Global Changes:

❖ It is very easy to specify a Global Change that will not change data in the way you intended.

❖ You must consider your Autocost rules when using Global Change on resources, percentages complete, and durations. For example, changing Original Durations will have no effect on the Early Finish of activities that have commenced when Remaining Duration and Percent Complete are unlinked.

❖ It is **STRONGLY** recommended that you always review the **Global Change Report** to review your changes before making permanent changes by running a Global Change.

❖ It is **STRONGLY** recommended that you consider making a copy of your project before using a Global Change: copy the project in the Enterprise Window, make a Baseline or use a Reflection Project and Change Report to review your changes before making permanent changes.

❖ When calculating Durations remember that P6 calculates in hours and if you are displaying durations in days, then you will need to divide or multiply as appropriate the durations by 8 to obtain the correct duration.

❖ A Global Change will only work on two variables and some calculations require more than two variables to achieve the required change. A **Temporary Value** may be created from the first two variable and the result stored in a **User Defined Field**. This **Temporary Value** may then be used on a subsequent line with the third variable. Any **User Defined Field** may be created and used as a Temporary Value.

❖ There are some functions that may be used with Global Change in the **Parameter/Value** field under **Then** and **Else** and that operate in a similar way to Excel. These functions may be used to populate User Defined Fields from other data fields as part of the process of editing Activity Descriptions and Activity IDs.

Global Change Function	Function Operation
DayOfWeek (Parameter)	Selects the weekday number of the date.
LeftString (Parameter,*)	Selects * of characters from the start of a field.
RightString (Parameter,*)	Selects * of characters from the end of a field.
SubString (Parameter,a,b)	From character "a" selects "b" number of characters.

25   MULTIPLE USER MANAGEMENT

25.1   Multiple User Data Display Issues

The following issues **MUST** be managed by the Database Administrator and have been covered in this publication in other sections:

❖ Any user, with access rights, may reset the database **Default Calendar** in the **Enterprise**, **Calendar** form, but this option will reset all users to the same calendar.

❖ By default, more than one person may open a project unless the **File**, **Open**, **Read Only** option is used or access is limited through **Security Profiles**. Thus, two people may make changes and create two versions of a project. Depending on who closes what and when, the final saved version may not be what it is thought to be. The **File**, **Refresh Data**, **F5**, option enables a user to refresh project data to see what other users changed.

❖ **User Baselines** are not **Project Baselines**. When a second user opens a project, which has a **Primary User Baseline** set by the first user, then this baseline will not be assigned to the second user. When the same layout is used to display the project, the **<Current Project> Baseline**, which displays the **Planned Dates**, will be displayed as the **Primary User Baseline**. Again, two users opening the same project and using the same Layout may display different data.

❖ It is possible to have two **Currencies** with the same symbol and if a user selects a different currency, then all costs displayed by the user will be converted to a different value. This option must be carefully monitored and if you do not need multiple currencies, then it is suggested that you should delete them all, to avoid any possible problems. If you are using multiple currencies, then make sure that all currencies have a different sign so there is no confusion.

❖ Users with different **Units Format** in their **User Preferences** will display different values for their units values which may be confusing when two users report two different resource values for the same project.

NOTE: It is critical for organizations to appoint a database administrator who understands these issues and keeps an eye on what is being sent to clients and makes sure that any display issues are either hidden or explained to the client in writing. Contractors may wish to consider making the system user and the project the same, as this resolves a number of issues. For example, User Filters and Layouts, including headers and footers, are by default the project's, reducing the possibility of sending out a report with the incorrect header or footer. User defaults become project defaults, resolving display issues. Access to the project may be easily restricted to the one user and therefore only one person may have the project open at one time.

25.2   Understanding F5

The author found that sometimes changes made in one place would not be reflected somewhere else; such as relationships added but not visible in columns, or WBS Nodes edited in the WBS Window but the changes not reflected in the Activities Window.

If this happens then select **File**, **Refresh Data** or press the **F5** key which:

❖ Writes your data to the database,

❖ Reads changes that other users may have made to your schedule,

❖ Puts the **Data Date** vertical line on the Gantt chart in the correct place when the system does refresh correctly and

❖ Corrects other display issues discussed above.

25.3   Understanding F10

The **File**, **Commit Changes** command writes any changes you have made to the database. These may then be read by other users by the **Refresh Data F5** command.

25.4   Preventing other users from changing my project

When a user opens a project, it is automatically opened as read/write and any user that has a project open may change the project data.

A project may only be opened as **Exclusive** (meaning that only the current user may edit it) by using the **File, Open...**form.

All other methods will result in the project being opened in the **Shared** mode and all users with access to the project may open and edit the project(s) at the same time. The **Shared** option may result in one user's edits being overwritten by another user's edits, depending on who saved the project and when.

A project that is opened in the **Shared** mode by multiple users with different **User Preferences Time Units** will result in the users calculating different values for Activity, WBS Nodes and Project durations in days.

26 MULTIPLE PROJECT SCHEDULING

26.1 Default Project

When multiple projects are opened together and each project has different **Schedule Options**, then the **Schedule Options** of all the projects will be changed and set to the same as the **Default Project** permanently without warning.

Version 20 allows the **Default Project** setting in the **Set Default Project** form to be overridden by selecting a different project in the **Scheduling Options** form, **Use scheduling options from** drop down box. This will not change the **Set Default Project** form **Default Project** but will override it and change all other projects **Scheduling Options** to the one selected in the **Scheduling Options**.

NOTE: If you are intending to open multiple projects together then it is best to ensure all projects have the same **Schedule Options**.

26.2 Multiple Projects in One Primavera Project

When there are many small projects that need to be managed, it would be logical to create a Primavera Project for each project.

On the other hand, one should also consider putting a number of small projects in one Primavera Project and have the projects identified by the first level WBS Node or some other coding, such as Activity Codes or Project Phase/WBS Category.

This is especially practical when there are many projects with a very small number of activities or when an organization only realizes benefits from a number of completed projects when they are all finished.

This option is also practical when one scheduler is managing all the small projects.

NOTE: The only problem with this approach is that P6 does not allow partial projects to be Baselined. This issue may be overcome by using User Defined Field Dates and a Global Change to set Baseline Dates for parts of a P6 project.

26.3 Multiple P6 Primavera Projects Representing One Project

Normally, one Primavera Project would be created for each of an organization's projects. There may be a requirement to break a Project down into Sub-projects, these reasons include:

❖ The project is large enough to require a number of schedulers and therefore a Primavera Project could be created for each scheduler to delineate each scheduler's area of responsibility.

❖ Two or more schedulers may open one project and access may be assigned down to WBS Node but the User Access has to be set up to allow them to be able to schedule and they are not able to link to other WBS Nodes.

❖ There could be a requirement to keep an individual organization's financial information confidential and as security and access is set at project level, information in one project may be hidden from specific users. This situation may exist when there are two or more contractors scheduling parts of a project and they require their cost to be kept confidential from other contractors.

❖ A project may have separable parts or multiple clients, but it is necessary to report the project parts separately, but allow resource management project wide. Again, a Primavera project could be created for each separate part of the project. In this situation each user may be given access to only one **Resource Node** from the **Global Access** tab of the **Admin**, **Users** form.

❖ A sub-project could be created as a Primavera Project for the security of sensitive financial information. The cost may be assigned to resources in the financial sub-project with access given to specific individuals. Activities in the financial sub-project may be LOE (Level of Effort) activities, spanning activities in other non-cost sub-projects. This method is generally suitable for high

© *Eastwood Harris*

level cost planning and management while allowing the detailed planning of a project in a non-financial sub-project without the burden of managing costs.

❖ When a Primavera Project is created for each sub-project, it would be logical to keep all the Primavera Projects located under one "project" EPS Node and assigned a single Project Code. All the Primavera Projects could be opened at one time for scheduling and reporting by selecting the EPS Node.

❖ The decision to break a project into two or more Primavera Projects must have a sound basis and be well thought out. The environment chosen should be well piloted and tested to ensure the desired results are obtained from the software. Planning and scheduling software is hard enough to use without adding the burden of creating multiple projects. There is a large amount of analysis that may be completed without using multiple projects. Filters may be used to isolate parts of a project and sub-net critical paths may be generated a number of ways, such as using the **Calculating Multiple Paths** function. You must ensure that the requirement to break a project into sub-projects using individual Primavera projects is well-founded.

❖ Some people suggest that sub-contractors should run their own sub-projects within a master schedule. My experience is that smaller or new sub-contractors often are very inexperienced at scheduling and many do not know the basics of scheduling. It is therefore unreasonable and risky to expect sub-contractors to drive strange and complex scheduling software and get it right. Some industries are better equipped to manage complex software, with skills found more likely in industries such as IT, but less likely to be found in the construction-related sector.

❖ It is also my experience that it is better to reduce the number of schedulers working on a project schedule to the absolute minimum required to manage a project. In large complex projects, these people need to be trained in the use of the software, be reasonably experienced running the software (or working under a person who is experienced) and run the schedule by an agreed-upon and documented set of guidelines.

❖ Managing the inclusion of sub-contractors' schedules always becomes an issue. Alliances tend to help resolve this problem as the schedules then become a joint responsibility.

26.4   Setting Up Primavera Projects as Sub-projects

There are a number of issues to be considered when moving to this environment. Be aware that Primavera does not have the sub-project options that are found in other products. For example, there are EPS Activity Codes, but there are no EPS Filters, Layouts, Resources or Schedule Options and a WBS may not be shared with more than one P6 project.

26.4.1   Opening One or More Projects

Enterprise and Project data may be accessed in the **Projects Window**. To access Project activity information, such as activities, resources, and relationships, a project must be opened and the **Activities Window** displayed. One or more projects may be opened at the same time by selecting one or more projects and/or selecting one or more EPS levels and then:

❖ Right-click and select **Open Project**,

❖ Select **Ctrl+O**,

❖ Select **File**, **Open…** to open the **Open Project** form:

The **Open** form enables the options of opening as **Exclusive**, **Shared** or **Read Only**.

NOTE: A project may only be opened as **Exclusive** (meaning that only the current user may edit it) by using the **Open Project** form. All other methods will result in the projects being opened in the **Shared** mode and all users with access to the project may open and edit the project(s) at the same time. The **Shared** option may result in one user's edits overwriting another user's edits, depending on who saved what and when. In addition, opening in the **Shared** option may result in different users seeing different values for Activity, WBS Nodes, and Project durations in days or hours if the users have different **User Preferences Time Units**.

26.4.2 Creating a WBS for Multiple Projects

It is not possible in P6 to share a WBS over multiple projects as P6 has limited "Sub Project" functions.

To have a single WBS over multiple projects one has to create an EPS or Global Activity Code and apply this to all the activities in the multiple P6 Projects.

26.4.3 Understanding the Default Project

When multiple projects are opened:

❖ The system selects the **Default Project** when two or more projects have been opened at the same time.

❖ The **Default Project Schedule Options** are used to calculate all the open projects.

❖ Select **Project, Set Default Project...** to open the **Set Default Project** form where you may change the default project:

Project ID	Project	Default	
EC00515	City Center Office Building Addition	☑	✓ OK
EC00501	Haitang Corporate Park	☐	⊘ Cancel
EC00610	Harbour Pointe Assisted Living Center	☐	
EC00620	Juniper Nursing Home	☐	❔ Help
EC00530	Nesbid Building Expansion	☐	
EC00630	Saratoga Senior Community	☐	

❖ All open projects **WILL** have their **Schedule Options** set to the same as the **Default Project** after the projects have been scheduled.

IMPORTANT NOTE: In Version 20 the **Default Project** setting in the **Set Default Project** form may be overridden by selecting a different project in the **Scheduling Options**

form, **Use scheduling options from** drop down box which will not change the selected project in the **Set Default Project** form but will override it and change all the other projects **Scheduling Options** to the one selected in the **Scheduling Options** form.

NOTE: When more than one Primavera project is opened at the same time and each project has different **Schedule Options**, then the non-default project's **Schedule Options** are changed to be the same as the default project's, without warning. These non-default projects may calculate differently when opened with other projects. In addition, the next time a non-default project is opened in isolation it may calculate very differently from the previous time it was opened in isolation. To prevent this, either all projects in each database must have the same **Schedule Options**, or access to projects carefully restricted, or ensure users only open one project at a time.

An example of changing the default project when each project has different options is demonstrated in the following picture. The first has retained logic and the second has progress override. Activity PG3-2 has moved forward in time as it is now being scheduled with Progress Override after initially being scheduled with Retained Logic. These types of unexpected changes may significantly affect your project and may occur when two or more projects, each with different Schedule Options, are opened together.

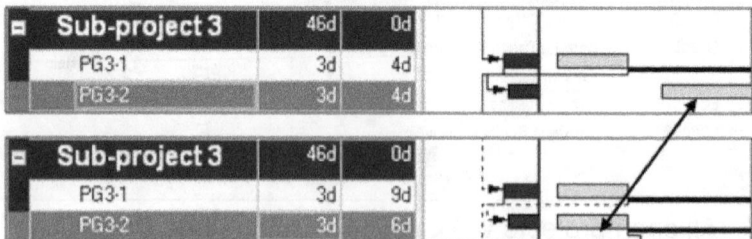

■	Sub-project 3	46d	0d	
	PG3-1	3d	4d	
	PG3-2	3d	4d	

■	Sub-project 3	46d	0d	
	PG3-1	3d	9d	
	PG3-2	3d	6d	

26.4.4 Multiple Projects with Different Data Dates – Version 19 and Earlier

The **Default Project** does not set the **Data Date** for all projects. In the example following:

❖ The **Default Project** t is **Data Date** 6 Mar and

❖ The other two projects have their **Data Date** set as per their names.

When scheduling, the following message is received:

Primavera P6 Professional R8.1

(?) Not all opened projects have the same Data Date. If you choose to continue, each project will be scheduled based on its own Data Date. Otherwise, you can set the Data Dates in the Projects View. Do you want to continue?

Yes   No

There are unlinked activities, MPS01-30, MPS02-30 and MPS03-30 in each project and it may be seen in the picture:

❖ All projects are scheduled with their own **Data Dates**,

❖ The **Data Date** line is shown on the earliest project **Data Date**, not the **Default Project Data Date**.

Activity ID	Original Duration	Total Flc	
Date Date 6 Mar			Default Project
MPS01-10	5d	0d	
MPS01-20	5d	0d	
MPS01-30	5d	5d	
Data Date 6 Feb			
MPS02-10	5d	0d	
MPS02-20	5d	0d	
MPS02-30	5d	35d	
Data Date 2 Jan			
MPS03-10	5d	0d	
MPS03-20	5d	0d	
MPS03-30	5d	70d	

The **Data Dates** of multiple projects may be set using a column in the **Projects Window** and utilizing the **Fill Down** function.

26.4.5 Data Date Selection in Multiple Project Scheduling – P6 Version 20 Enhancement

Version 20 added the additional option of:

❖ **All Projects use their own data dates** which is how P6 Version 19 and earlier versions calculated, or

❖ **Apply selected data date to all open projects** and all projects will have their **Data Date** permanently changed to the selected date.

When scheduling multiple projects you now have a choice of the **Data Date** used for calculating:

❖ All projects may be assigned a new **Data Date** in the **Schedule** form, or

❖ New in P6 Version 20, you may schedule all projects with their own **Data Date**.

26.4.6 Multiple Project Scheduling Options Selection– P6 Version 20 Enhancement

In the P6 Version 19 and earlier when multiple projects were scheduled together and they had different **Scheduling Options**, which could happen when users change them from default or when you imported a project, then all the different **Scheduling Options** of all the projects being scheduled were changed to the **Default Project** on a permanent basis and changed projects would calculate differently from then on.

The Default Project may still be set by selecting **Project**, **Set Default Project**:

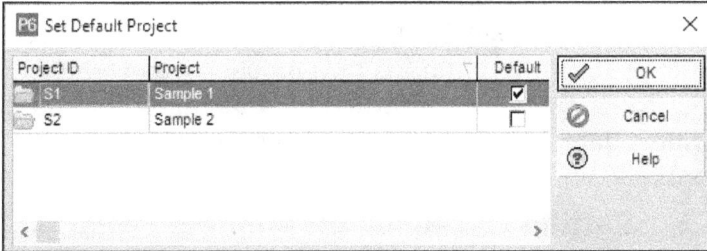

But now you may select which project **Scheduling Options** are used when scheduling multiple project in the **Scheduling Options** form which is opened by selecting **Tools**, **Schedule**, **Options**:

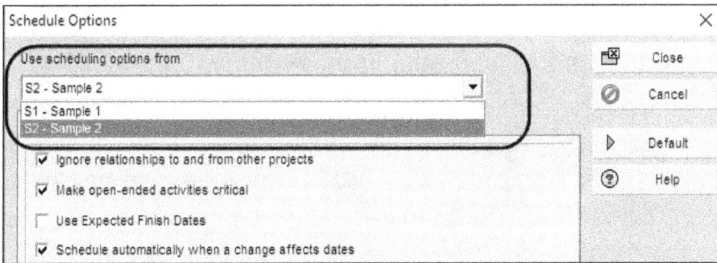

What happens now?

❖ The **Default Project** is ignored,

❖ The **Scheduling Options** from the project selected in the **Scheduling Options** are used to calculate all the project, and

❖ The other project(s) **Scheduling Options** are changed permanently to the project selected the **Scheduling Options** form on a permanent basis.

NOTE: Thus, the problem of **Scheduling Options** being changed when scheduling multiple projects has **NOT** been solved. You now have two options to mess up your **Scheduling Options** of projects when scheduling multiple projects with different options.

Again, I reiterate you should make all the **Scheduling Options** in one database the same when you schedule multiple projects and be careful when importing projects.

26.4.7 Total Float Calculation

In P6.1 and earlier versions the Total Float of each project is calculated to the last activity of each individual project schedule. In Primavera 6.2 a new function was created under **Tools**, **Schedule...**, **Options...**, **Calculate float based on finish date of** which resolves this problem. More information in para 16.11.

26.5 Setting Baselines for Multiple Projects

Baselines may be created for all the multiple P6 projects using the **Maintain Baselines** form. There are two options:

❖ Open the **Maintain Baselines** form by selecting **Project**, **Maintain Baselines...**,

❖ Select [⊕ Add] to open the **Add New Baseline** form and create the new baselines, either you may:

➢ Use the option **Save a copy of the current project as a new baseline**.
   NOTE: This process is **NOT** recommended because **Ghost Relationships** are created when you restore any multiple project baseline as, please read the next paragraph to understand why, or

➢ Make a copy the projects in the **Projects Windows** and then you may use the **Convert another project to a new baseline of the current project** to create baselines.
   NOTE: This is the recommended option.

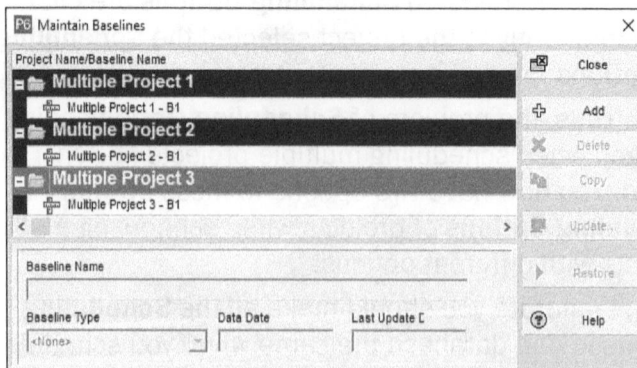

❖ Select **P**roject, **Assign Baselines…** to open the **Assign Baselines** form and select one project at a time to assign the baselines.

Assign Baselines		✕
Project	✓	OK
MP1 : Multiple Project 1 ▾	⊘	Cancel
MP1 : Multiple Project 1		
MP2 : Multiple Project 2	?	Help
MP3 : Multiple Project 3		
<Current Project> ▾		
User Baselines		
Primary		
<Current Project> ▾		
Secondary		
<None> ▾		
Tertiary		
<None> ▾		

NOTE: Remember, a **User Baseline** set by one user will not be displayed when another user opens the project. The **<Current Project> Baseline** displays the **Planned Dates** from the current schedule and will be shown as a baseline

26.6   Restoring Baselines for Multiple Projects

Schedulers often wish to restore baselines to inspect or review the original schedule.

NOTE: The process using **Save a copy of the current project as a new baseline** identified in the previous page results in one interesting issue when Baseline projects are restored. The software creates **Ghost Relationships** between the Current Schedules and Baseline schedules which must be avoided because there is a high risk that neither the Baseline nor Current projects will calculate correctly once Multiple Project Baselines are restored.

The example below explains what happens when three simple projects are baselined together:

❖ The three projects were opened together and baselined:

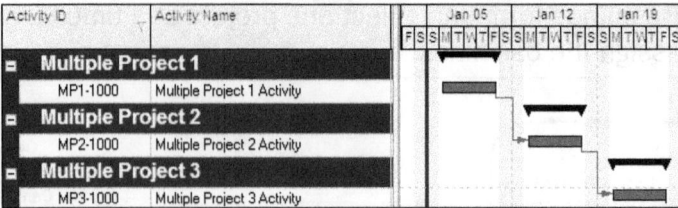

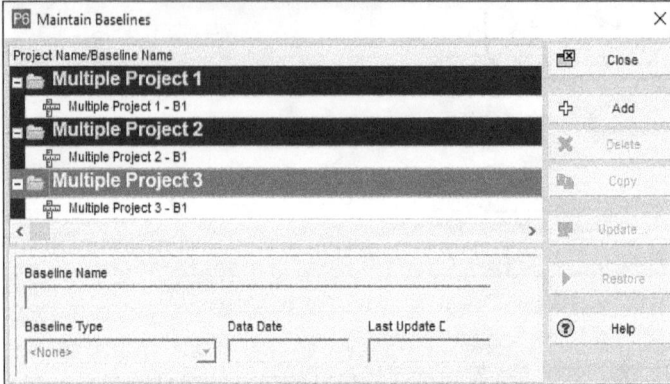

❖ The baselines were unlinked and restored,

❖ When the current and baseline projects are opened together there are unwanted **Ghost Relationships** created by the software, without warning, between the Current and Baseline projects:

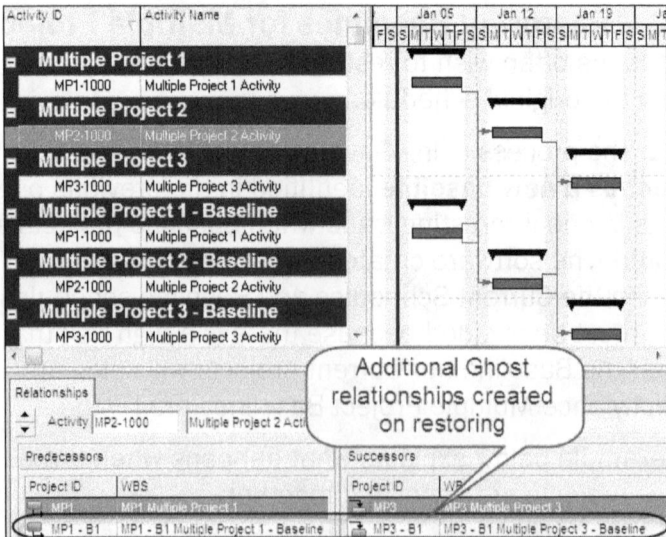

Additional Ghost relationships created on restoring

❖ When the current projects are opened on their own and activity durations shortened you will see that the schedule does not calculate correctly because of the **Ghost Relationships**:

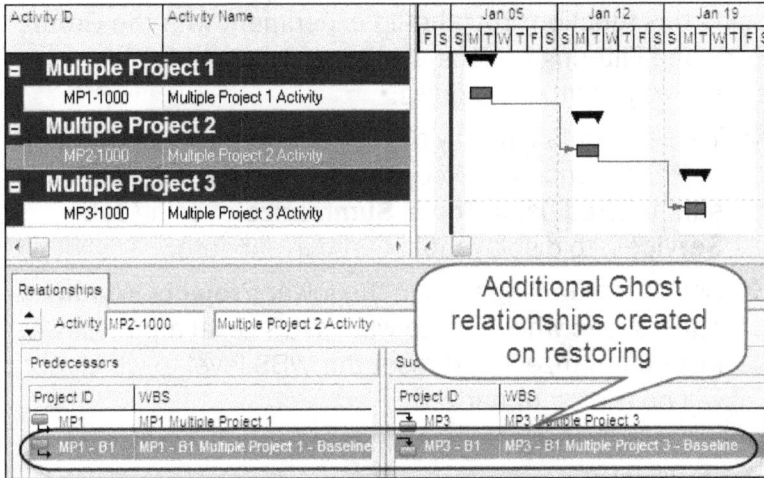

Therefore, if you wish your baseline projects to maintain the relationships to other baselined projects only and not have **Ghost Relationships** created with the current projects when the multiple current projects with relationships amongst them are baselined, then you must:

❖ Open the **Projects Window**,

❖ Copy the multiple projects in this view **Projects Window**,

❖ Then set the baselines using the **Convert another project to a new baseline** of the current project option in the **Maintain Baseline** form.

Now if the baselined projects are restored there will be no **Ghost Relationships** created.

26.7 Tracking Window

Tracking Layouts are used for the resource, cost, and schedule analysis of multiple projects. This section introduces the concepts but does not go into the detail of using this function. You should experiment with the Group, Sort, and Filtering options available, which all function in a similar way to other windows.

❖ These layouts typically display summarized data to EPS or Project and WBS Node level. The data must be summarized using **Tools**, **Summarize** or using **Job Services**, to display the latest current data.

❖ Select the **Settings** tab in the lower **Projects Window** pane to see when and to what WBS Level a project was last summarized, and reset the WBS level to which data will be next summarized.

There are four **Tracking Layout** types and a new layout is created by:

❖ Saving an existing layout, saving with a new name and editing it, or

❖ Selecting **View, Layout, New Layout...** which opens the **New Layout** form:

New Layout	
Layout Name	✓ OK
OzBuild resources	
Available to	⊘ Cancel
Current User ▼	
	⑦ Help
Select Display Type	
⊞ ○ Project Table	
⊟ ○ Project Bar Chart	
⊪ ● Project Gantt/Profile	
⊪ ○ Resource Analysis	

The following pictures indicate the type of data a **Tracking Layout** will display:

❖ **Project Tables** display columns of data for selected Projects or WBS Nodes:

❖ **Project Bar Charts** display the selected projects WBS Node data in horizontal bars:

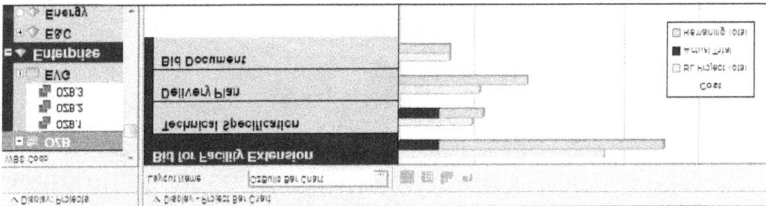

❖ **Project Gantt/Profiles** display three panes, with bars in the top right pane and either a spreadsheet or a profile in the bottom pane.

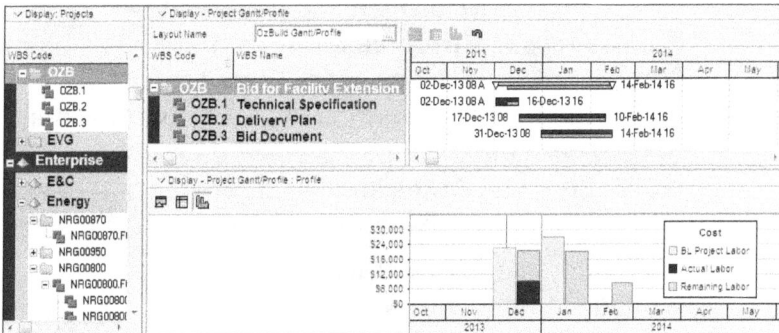

❖ **Resource Analysis** displays four panes:

  ➤ The projects to be analyzed are selected in the top left-hand pane,

  ➤ The resources to be analyzed are selected in the bottom left-hand pane,

  ➤ Bars, Resource Profiles or a Resource Table may be displayed in the top right-hand pane, and

➢ The bottom right-hand pane may display either a Resource Profile or a Resource Table:

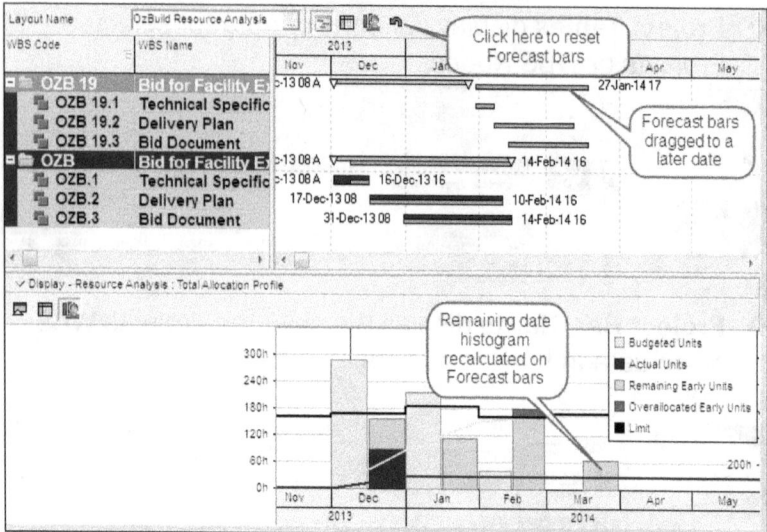

➢ An existing layout may be seen by opening the **Open Layout** form. Select **View, Layout** or click the ⬛ icon in the top right-hand pane.

➢ The **Forecast Bars** in the picture above have been dragged to a new location and the **Edit**, **User Preferences...**, **Resource Analysis**, **Time-Distributed Data** option set to **Forecast dates**, allows the **Resource Remaining Early units/costs** to be recalculated on the **Forecast dates**.
NOTE: Forecast Dates are a separate set of dates allowing "what if" scenario planning.

➢ The bottom pane of a Tracking Layout may be hidden, as with other windows.

➢ You should experiment by right-clicking in all the panes to see all the display options.

27 Exporting and Importing

P6 has functions available for importing and exporting data found under the file command. These are fairly simple to use, but have a high risk of corrupting your database if you do not understand the options.

If they fail to operate, then the first thing to confirm is that your software is patched with the latest patches.

Read the Administrators Guide carefully before importing a project, as this is a very complex operation and may import unwanted data into your database.

When projects are imported or exported to other scheduling packages, they will often calculate differently due to the different methods of calculation of each package. Do not expect to import from Microsoft Project or any other software and expect to see the same dates when scheduling.

There are some articles on **www.primavera.com.au** and **www.eh.com.au** that explain the issues.

Importing a file from another Primavera database may give different results depending on the database and **User Preferences** and **Admin Preferences** in each database, and these should be carefully checked.

Importing a project into a working P6 database must be carefully planned to ensure that existing projects are not impacted by the imported data, and the options available on the import wizard are fully understood. Updating or overwriting existing data may affect existing schedules.

It is recommended that you establish a sacrificial database into which you import projects, so that corporate databases are not filled up with unwanted data.

You may wish to invest in one of the third party tools listed on www.primavera.com.au to clean up you data before importing a project.

27.1 Supported Project File Types

The following file types that may be used to import and export complete projects are:

❖ **XER** – Used to exchange one or more projects between Primavera databases, regardless of the database type in which it was created and exports all project data. Earlier versions of **XER** files may be imported into later version databases.

NOTE: A layout (formatting) is not part of a **XER** file. Baselines are not exported with a **XER**. The web tool does not import or export **XER** file. It is better to use P6 **XML** to export and import P6 projects.

❖ **XML** – A format introduced with Primavera Version 6.0 which is used to import data from the Project Manager module. This is the same software language, but a different format to a Microsoft Project XML file. It has the ability to import and export Baseline projects and Project Layouts.

NOTE: After importing an XML you have to recalculate if you wish to see Float values and the critical path. Also an XML file will recalculate the resource costs when the resource rate has been changed after the resources has been assigned. This has caused issues for contractors submitting projects in XML format to clients as the client's version has different costs to the contractor version. P6 Version change how the resource cost for imported costs but did not resolve the issue. I thus recommend that a resourced schedule should be submitted to a client in XER format. See my paper on my web sites titled **IMPORTED P6 PROJECTS CALCULATING INCORRECT COSTS**.

❖ Microsoft Project **XML** format is supported in Version 6.2 and later. This allows import of a file created by Microsoft Project 2000 and later without the installation of Microsoft Project. The file needs to be saved from Microsoft Project in XML format.

❖ **UN/CEFACT XML** format. P6 Version 8.3 introduced the support of the format from the **File**, **Export** menu. This format is mandated by many US Government agencies.

❖ **IPMDAR** format introduced in P6 Version 21; an **IPMDAR** is a project export format required by the US Department of Defence from 2021. It is only available when you open an EEM database, when exporting from a PPM database the option is greyed out:

❖ **P3** and **SureTrak** are very old Primavera software packages and will not be discussed

❖ ***.MPP** This is the default file format that Microsoft Project uses to create and save files. There are three different formats Microsoft Project 2010 – 2019, Microsoft Project 2007 and Microsoft Project 2000 – 2003. Earlier versions of Primavera P6 will not import any **MPP** file when Microsoft Project 2007 or later was installed, as these disable the **MPP** import function. The import and export of **MPP** files in later versions of Microsoft Project is no longer supported by P6. Microsoft Project files should be saved in Microsoft XML format for importing and exporting.

❖ ***.MPX** is a very old text format created by Microsoft Project 98 and earlier versions and will not be discussed. It should not be used as it strips calendars from activities.

NOTE: When saving to an XML file in either P6 of MSP it is good practice to add MSP or P6 into the file name, so you know the format of the file, as P6 and MSP XML files have a different format.

27.2   Other Data Files Types

The following are a summary of other files types:

❖ **PLF** – Used to exchange **Layouts** between Primavera databases, regardless of the database type in which it was created. In Primavera Version 15.1 PLF files may now be imported into Visualizer.

❖ **ANP** – Used to save the position of activities in an **Activity Network**.

❖ **ERP** – Used to exchange **Reports** between Primavera databases, regardless of the database type in which it was created.

❖ **PCF** – Used to exchange **Global Changes** between Primavera databases.

❖ **VLF** - Visualizer Layout File allows the import and export of Visualizer Layouts between users.

❖ **XLS**. Primavera Version 5.0 introduced a function allowing the import and export of data in Excel format. It is best to export some sample data first and then update the data and import it.

27.3 Send Project

The **File, Send Project...** should create a **XER** file (an export file) automatically and attach it as an attachment to an email.

This requires P6 and the email software to be loaded on the same machine and will not work in cloud or virtual environments.

27.4 Unable to import a XER created in a later version of P6.

To be able to import a XER file created from a later version of P6 the XER file should be opened with **Notepad** and the version changed:

```
   EC00501.xer - Notepad

 File  Edit  Format  View  Help
ERMHDR  21.12   2022-05-30       Project admin    admin    dbxData
%T      COSTTYPE
%F      cost_type_id    seq_num cost_type
%R      4       50000           terials
%R      478     10001     Equ     e
%T      CURRTYPE                          Edit the P6 Version
%F      curr_id decimal_digit_c           number here          imal
%R      1       2       $                                       .1)
%R      13      2       £           .     ,       #1.1    (#1.1)
%R      14      2       ¥           .     ,       #1.1    (#1.1)
```

27.5 Importing from Excel

Some data associated with an imported activity must exist before the activity is imported from Excel, otherwise it will not be imported. This includes items such as Roles, Resources, and Activity Codes. In older versions of P6 these data items could be imported using the Primavera SDK (which is loaded from the installation CD and instructions are available on the Administration Guide) and an Excel spreadsheet. Guidance is available from the Oracle Primavera Knowledgebase. This function is no longer available with later versions of P6, but there are some third-party software packages that will do this. A list of these software packages is maintained **at www.primavera .com.au**.

27.5.1   Notes and/or Restrictions on Export

A few points to understand when using the Primavera Excel Import function:

❖ The following sheets are created on export and these sheet names must not be changed:

➢ **TASK** containing Activity data

➢ **TASKPRED** containing Activity Relationships data

➢ **PROJCOST** containing Expenses data

➢ **RSRC** containing Resources data

➢ **TASKRSRC** containing Resource Assignments data

➢ **USERDATA** containing user data that should not be changed.

❖ Do not change the language between importing and exporting.

❖ The first row of data in each sheet that is exported contains the database field name. The first row must not be changed otherwise the data will not be imported.

❖ The second row in the spreadsheet contains **Captions** that are deleted on spreadsheet import by the **"Delete This Row"** entry in the right column of the spread sheet. This **"Delete This Row"** entry may be copied to a line of data that is to be deleted from the project on import.

❖ Dictionary data, such as Activity Codes being imported, must exist before the data is imported.

❖ Only Activity Codes may be imported. If you wish to import the Activity Code descriptions, then you will have to use the Software Developers Kit (SDK) in P6 Version 8.3 or earlier or a third party tool.

❖ Only a maximum of 200 columns of data may be exported.

❖ **Sub-units** of time are not supported and the Sub-unit check boxes in the **Edit, User Preferences..., Time Units** tab should be unchecked.

❖ **Percent Complete** must be a value of between 0 and 100.

❖ Anything listed as a field may be exported.

❖ The **User Preferences** will affect how your data is exported and may give different values for resources.

27.5.2 Notes and Restrictions on Import

When attempting to import data using this type of tool there are some guidelines that apply to many applications, not just to this Primavera tool:

❖ Create a test project and experiment with this function before using it on a live project.

❖ Export some data first as this exports the correct column headings and sheet names.

❖ Change or add data to the exported spreadsheet and import new data into the test environment. Then review that the data is importing correctly and that the schedule is calculating as expected.

❖ Back up or take a copy of your live project before importing into a live project.

❖ It is often better to import into User Defined Fields to ensure the data gets into the database and then Global Change into the desired place.

❖ Activity data must have the Activity ID and WBS Code, as these are the unique identifiers for each activity within a database.

❖ The **delete_record_flag**, in the far-right hand column, titled **Delete titled this row** against line 2 of the Excel spreadsheet deletes the line 2 activity on import.

❖ The **Delete This Row** flag may be placed against any spreadsheet line and the activity will be deleted on import.

❖ **NOTE:** The **Delete This Row** flag did not delete rows in the authors P6 Version 21.

❖ Calculated fields may not be imported and are marked with an (*), see picture below:

	D	E	F
	task_name	start_date	end_date
de	Activity Name	(*)Start	(*)Finish
	Identify Supplier Components	11/12/2013 8:00:00 AM	12/12/2013 4:00:00
		13/12/2013 8:00:00 AM	16/12/2013 4:00:00
		2/12/2013 8:00:00 AM	
		2/12/2013 8:00:00 AM	4/12/2013 4:00:00 F
		4/12/2013 8:00:00 AM	10/12/2013 4:00:00

These may not be imported

❖ To see if the data field you wish to import may be imported, export the field and see if the field has an (*) by the second line description in the spreadsheet. Fields that may not be imported include, but are not limited to:

➢ Most dates except the Actual Start and Actual Finish,

➢ Expected Finish,

➢ Actual, Remaining, and At Completion Durations.

❖ Therefore, if you wish to import dates to create un-started activities without importing the Original Duration, then you will have to import the activity with **Actual Start** and **Actual Finish** dates where you want the activity to lie and use a Global Change to take-off the Actual Dates:

	A	B	C	D
1	task_code	wbs_id	task_name	act_start_date
2	Activity ID	WBS Code	Activity Name	Actual Start
3	OZ1140	OZB.3	New Activity	5/01/2014 8:00:00 AM

Then	Parameter	Is	Parameter/Value	Operator	Parameter/Value
	Original Duration	=	Actual Duration		
And	Primary Constraint	=	Start On		
And	Primary Constraint Dat	=	Actual Start		
And	Actual Start	=			
And	Actual Finish	=			

❖ When only exporting some data on an occasional basis, then it may be easier just to copy and paste the data into a spreadsheet.

If you wish to bring across the band headings into Excel, then the Activity ID must be displayed in the first column.

There is more information in the Help file under **Reference**, **Importing and Exporting**.

27.5.3 Importing and Exporting to Microsoft Project

The P6 routine to import and export a Microsoft Project file has been modified over time and does not always operate as one would expect, but some versions have displayed interesting features including:

❖ Projects with blank lines will not import.

❖ Project with a "/" in fields such as Resource Names not importing.

❖ In some P6 Version the Microsoft Project 24 hour per day calendars imported as 1 hour per day in P6 and give gross calculation differences.

❖ Milestones in Microsoft Project are calculated as Finish Milestones but are imported into P6 as Start Milestones and give calculation differences.

❖ Activity Spits are not supported in P6 and should be removed before importing.

❖ As Late as Possible in P6 is calculated as Zero Free Float and as Zero Total Float in Microsoft Project, and give calculation differences.

❖ A Microsoft Project file is imported into P6 with a Project Must Finish by constraint, this usually has to be removed.

❖ A P6 baseline is a complete project, Microsoft Project just records the Start, Finish, Duration, Work and Costs.

Remember P6 has more scheduling functionality than Microsoft Project and not all data may be imported, and it is better if these additional functions in P6 are not used:

❖ P6 allows 4 relationships between two activities and Microsoft Project only one,

❖ P6 allows two constraints per activity and Microsoft Project only one,

NOTE: After importing a project from one product to another, then cut and paste them both into Excel and compare them.

27.6   Importing from another P6 Database

As discussed earlier:

❖ Importing a file from another Primavera database may give different results depending on the database and user preferences in each database, and these should be carefully checked.

❖ Importing a project into a working P6 database must be carefully planned to ensure that existing projects are not impacted by the imported data, and the options available on the import wizard are fully understood. Updating or overwriting existing data may affect existing schedules.

NOTE: External Dates will replace links to other projects in the source database and then act like Early Start and Late Finish constraints, and will affect the schedule calculation. You should always check for External Dates when importing a project.

27.7   Import and Export Filters

It is not possible to import and export a filter in isolation, but if a filter is made **a Layout Filter** it may be exported as part of a layout.

27.8   Import or Export data with Keystroke recorders

To bring into P6 large amounts of data from products that do not Export, such as maintenance management systems that do not export data easily, then you may consider investing in a keystroke recorder, of which there are many on the internet. To use a keystroke recorder:

❖ Locate, buy, install and learn the software,

❖ Record one operation, such as creating an activity and then copying the description from a maintenance management program into P6.

❖ Then run the routine as many times as you require to create multiple activities.

27.9 Importing Activity Codes into P6 using and XER file and Excel

The Primavera P6 Excel spreadsheet created when you export a P6 project does not allow the creation of Activity Codes and their description on import. These have to exist before import.

To import a large number of Activity Codes that exist in Excel or a text document, you may edit a XER file using Excel:

❖ Create a P6 Project,

❖ Add an activity so there is activity data to export,

❖ Create a **Project Activity Code** in the **Activity Code Definition** form,

❖ Create one **Activity Code Value** and **Activity Code Description**,

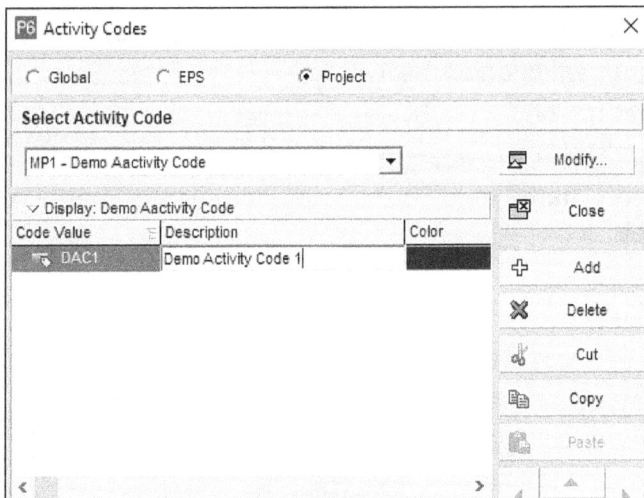

❖ Export the project as an XER file,

❖ Change the xer file extension to *.txt,

❖ Open the renamed *.xer file with Excel,

© *Eastwood Harris*

❖ You should be asked to Data Parse the file and accept the defaults:

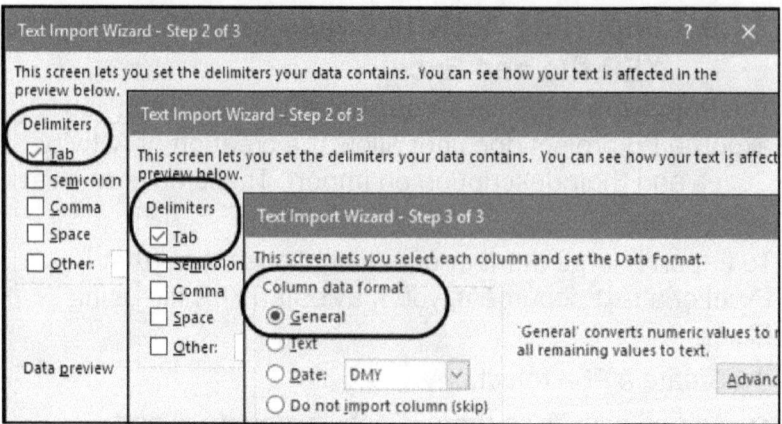

❖ Find the **ACTVCODE** area:

	A	B	C	D	E	F
37	%T	ACTVCODE				
38	%F	actv_code_id	parent_ac	actv_code_type_id	actv_code_name	short_name
39	%R	7351		1568	Demo Activity Code 1	DAC1
40	%E					

❖ Insert the rows into the spreadsheet for the Activity Code Vales and Activity Code Descriptions,

❖ Add the new data by copy and paste,

❖ Increment the **actv_code_id**, **actv_code_name** and **short_name** fields but do not increment the **actv_code** field:

	A	B	C	D	E	F
37	%T	ACTVCODE				
38	%F	actv_code_id	parent_ac	actv_code	actv_code_name	short_name
39	%R	7351		1568	Demo Activity Code 1	DAC1
40	%R	7352		1568	Demo Activity Code 2	DAC2
41	%R	7353		1568	Demo Activity Code 3	DAC3
42	%R	7354		1568	Demo Activity Code 4	DAC4
43	%R	7355		1568	Demo Activity Code 5	DAC5
44	%R	7356		1568	Demo Activity Code 6	DAC6
45	%E					

❖ Do not forget to fill down the **seq_num** and **color**, not shown in the picture above,

❖ Save Excel worksheet,

❖ Change the file extension *.XER.

❖ Import the XER file into Primavera P6:

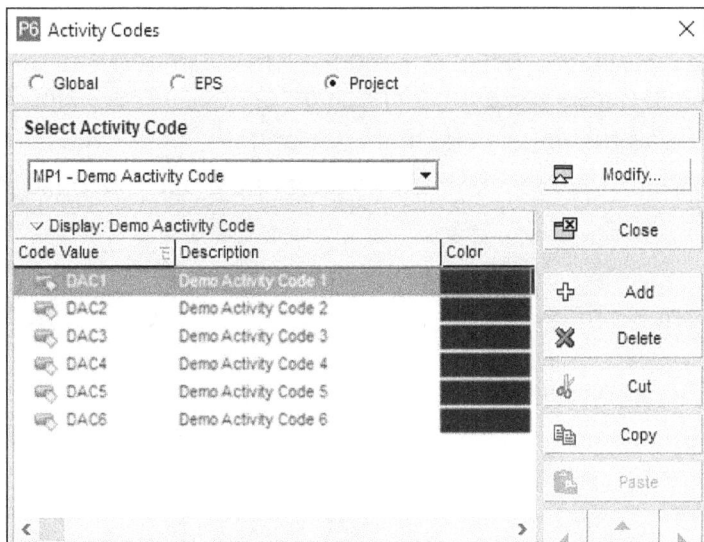

NOTE: If you are not careful, this is a very quick way of corrupting your database.

27.10 Lean Tasks may be imported using an XML file

Primavera Cloud allows activities to be broken down into tasks. **Lean Tasks** are similar to the P6 **Steps** function but with more functionality such as allowing the assignment of tasks to subcontractors. Lean Tasks may have the following attributes:

❖ Assignment of a duration,

❖ May be made "Private" so other people do not see them,

❖ Tasks may have logic between them which are called "Handoffs",

❖ A Task's Float is called "Slack".

❖ The planning of Tasks is completed on a **Planning Board** that may be shared between team members. This is a tool designed for short term planning by the people completing the work.

When you are using an EEPM database you may Import Lean Tasks through an XML file from Primavera Cloud.

27.11 Why will my import not work?

There are many reasons why a file will not import and there are two things you should check before calling for help:

❖ Read the import log for errors, and

❖ Check that your version has not had a patch release to solve an import issue.

28   Other Tools and Features

28.1   Check In and Check Out

The **Check In** and **Check Out** functions enables a project to be copied from a database, worked on in a remote location such as a client's database, and then be checked in to the original database at a later date, and the original schedule updated with the changes.

The file format of a **Checked Out** file is the same as a project exported in XER format, but checking out a project places a **Read Only** attribute on the project, and then it may be opened but not edited.

NOTE: External Dates will replace links to other project in the source database and then act like Early Start and Late Finish constraints and will affect the schedule calculation. You should always check for External Dates when importing a project.

28.2   Activity Discussion Feature

A new tab has been added to the **Activities Window** in P6 Version 8.3, titled **Discussion** which enables:

❖ Users to create a discussion thread for each activity,

❖ Each entry is saved with the date and user name and the entire thread is recorded, allowing an interactive discussion between users who have access to P6.

28.3   Advanced Schedule Options

The **Float Path** function enables individual critical paths to be banded as in the following picture and is useful when analyzing larger projects that have more than one critical path.

This is similar to Grouping by Total Float, but this function numbers the Paths, and each path contains activities that are linked, whereas banding by Total Float may group unlinked activities.

The following schedule has three chains of events on the critical path:

❖ Grouped by **Total Float**: and the chains are unclear:

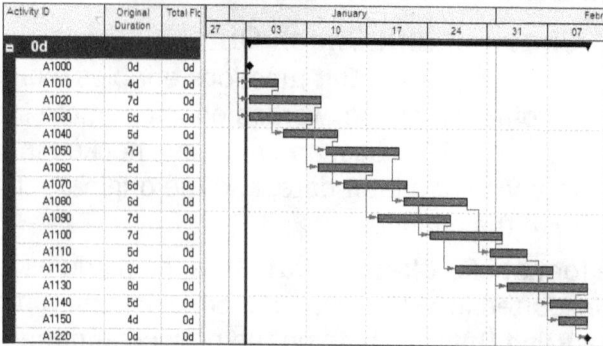

❖ Grouped by **Float Path** and the chains are clear:

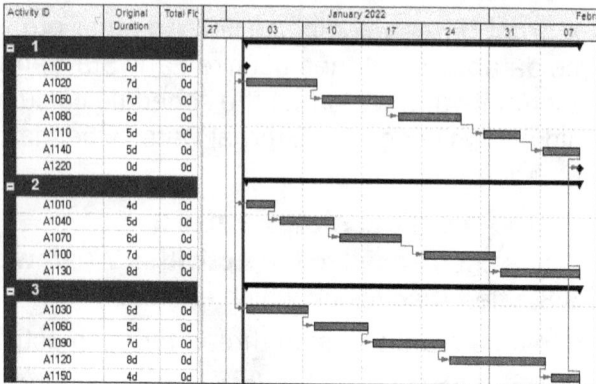

There are two steps involved, firstly calculating the multiple paths and secondly displaying the multiple paths.

28.4   Reflection Projects

Primavera Version 6.0 introduced a Reflection project function. A Reflection is a "What-if" copy of a project that may be edited and then merged back into the original project. The changes that are required to be kept may be incorporated into the original project and those not required may be ignored.

The **Reflection project** may be shared with a wider audience and people asked to view and make changes to the project. The **Reflection project** may be exported and sent to a customer who may make changes and then imported back into the database.

28.5 Who Has the Project Open?

When a project is opened with Primavera using the **File**, **Open** option, the **Open Project** form has **Access Mode** options to open the project as **Exclusive**, **Shared**, or **Read Only**.

28.6 Deleting Mass Data from a Project

When you want to delete a large amount of data from a project before giving it to a client or contractor, such as:

❖ All the **Notes**

❖ All the **Costs**

❖ All the **Resources**

Then there are two tools to consider:

❖ Display the appropriate column and use the **Fill Down** function, and

❖ Select the project in the **Project Windows** and copy and paste, and uncheck the data types you do not want to be copied.

28.7 Managing P3 and SureTrak files

If you have a later Windows machine, then you will not be able to load P3 or SureTrak. To be able to open these old files, I maintain a Windows XP Virtual Machine with SureTrak, P3 and an old version of P6 loaded, to convert files from P3 or SuraTrak to XER format.

28.8 Finding a Bar in the Gantt Chart

If your bars are not in view in the Gantt Chart, then just double click in the Gantt Chart, in line with the activity and the bar will be brought to the mouse pointer.

28.9 Check for Open Ends

An Open End is an activity without a successor.

❖ One way to check for **Open Ends** in a schedule is to display the log file after scheduling. To do this select **F9** or **Tools, Schedule, View Log.**

❖ You may also display the **Successor** column and search or filter for blank cells, or in later versions of P6 display the **Successor Details** column, which will also display the relationship type and lag.

28.10 Chasing or Tracing Logic

Often people wish to identify a chain of linked activities to check the logic. Here are some ideas for you to consider:

❖ The **Network Diagram** allows the selection of two non-adjacent WBS Nodes and will display the relationships between two or more selected nodes.

❖ The **Trace Logic** function may be used and is able to be formatted in the Trace Logic Options to display the desired number of Predecessor or successor Levels:

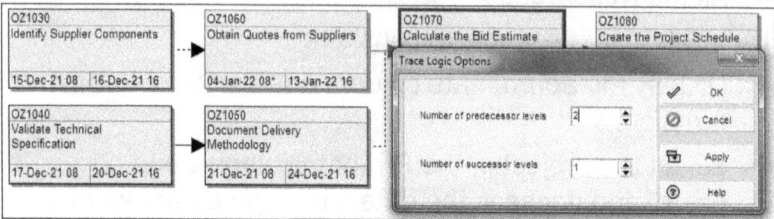

❖ Display the first activity only with a **Filter** and then display the **Predecessor** or **Successor** form in the lower pane and use the **GoTo** [GoTo] button to display the chain of activities.

❖ The **Tools**, **Schedule**, **Advance** form allows the calculating of **Multiple Critical Paths**, and then the activities are Grouped by **Float Path** to see chains of activities grouped by their **Float Path**.

28.11 Why have I got Negative Float?

Negative Float is created several ways:

➢ Applying a **Project Must Finish By Date** earlier than the latest activity finish date,

➢ Applying a constraint such as a **Finish No Later Than**.

➢ Linking to an unopened project.

➢ Importing a project that has **External Early Start** or **External Late Finish** dates created by exporting a P6 project that has been exported from a database that has links to other projects that have not been exported. These dates represent the missing links and work in the same way as constraints.

➢ Putting in logic that confuses P6. If you change a calendar between an Activity and Finish Milestone you will get Negative Float, but when you change the milestone to a Start Milestone and schedule, the Negative Float goes away:

Calendar	Activity Type	Total Float	Original Duration	Sun	Mon	Tue	Wed	Thr	Fri	Sat	Sun	Mon
							Jan 15					
5x10	Task Dependent	-3d	5d									
7x24	Task Dependent	-2d 12h	2d 12h									
5x10	Finish Milestone	0d	0d									
5x10	Task Dependent	0d	5d									
7x24	Task Dependent	0d	2d 12h									
5x10	Start Milestone	0d	0d									

28.12 Causes of Rounding Errors in P6

You may find that your activities are not starting at the start of the day or ending at the end of a day, or a one-day activity is spanning two days etc.

I teach in all my courses the user must always display:

❖ The time with the date and

❖ Durations with the subunits or decimal unit.

These are set in the **User Preferences** and these issues are then isolated quickly.

Firstly, it is important to understand that P6 rounds down values of *.5, so 8.5 is rounded to 8 and 8.51 is rounded to 9.

The possible reasons for rounding errors or display issues are:

❖ The activity **Duration** is not a round figure and the display has been rounded down, but you are not showing decimal days or days and hours settings in the **User Preferences**.

❖ The **Remaining Duration** is not a round figure and the display has been rounded down. This is caused by using **Duration % Complete**.

❖ The calendars do not have the same number of hours per day for every day in a calendar.

❖ The calendar **Hours per Time Period** do not match the calendar hours per day.

❖ The activity is not starting at the start of the day. Check if there is a change of calendar from one activity to another. One way to stop this happening is to have all calendars in a multiple calendar schedule ending at the same time.

❖ The **Project Start Date** is not aligned with the calendars. For example, the project start is at 08:00hrs but the first activity calendar is set to start at 07:00hrs.

28.13 Store Period Performance – P6 Version 19 and Earlier

This function stores the actuals for each period in dedicated fields for each financial period. There are several reasons for doing this:

❖ In order that the amount spent in each period is recorded, and

❖ Histograms and S-Curves will not change each period when the spend is not linear, resulting in more accurate reports.

This function is very difficult to use, and setup takes place in several steps:

❖ The periods are set up in **Admin, Financial Periods,**

❖ The **Calculations** tab in the **Projects Window** option **Link Actual to date and Actual This Period Units and Cost** must be checked,

❖ The period data is stored after each schedule update using **Tools, Store Period Performance....**

❖ The **Edit, User Preferences, Application** tab, **Columns** section, **Select financial periods to view in columns** enables the user to restrict the number of columns that are displayed in forms such as the **Columns** form, thus reducing the amount of scrolling required to find a specific column.

❖ Finally, these results may be viewed and edited in the **Past Period Actuals** columns of the **Resources Assignments Window, Activity Details Resources** tab, the Activity Table, etc.

❖ **The Link actual to date and actual this period units and costs** sometimes has to be unchecked to allow editing of values when data has been incorrectly entered for past periods.

❖ The options to display **Financial Period** values is available in forms like the **Activity Usage Profile Options.**

28.14 Different Projects may have different Financial Periods – P6 Version 20 Enhancement

You may now assign different Financial Period to projects in P6 Version 20 by the use of **Financial Period Calendars**. When you select **Admin, Financial Period Calendars** you may create multiple **Financial Period Calendars** and each may have different Financial Periods:

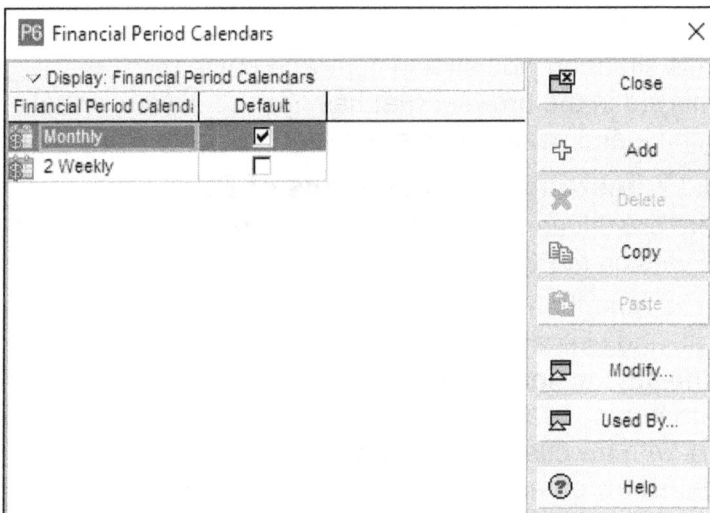

The **Financial Period Calendar** is assigned to a Project in:

❖ The Projects Window, General tab, or

❖ By displaying the Financial Period Calendar column:

NOTE: This function did not work in the authors P6 Version 20.12.0.37740, but updates should fix this issue. As an interim you may change the **Default Financial Period Calendar** before creating a project.

28.15 What are the differences in P6 Versions?

Oracle Technology Network website has an Excel spreadsheet called the **Cumulative Feature Tool** which displays changes in versions and bug fixes.

28.16 I am on a Virtual Machine and Help will not load?

In this situation, you should copy and paste the URL displayed in the browser that has opened in your virtual machine and paste it into a browser that does work.

28.17 Installing two versions of P6

There are documents available that explain how to install two version of P6 on one machine.

On the other hand, I prefer not to install P6 on my machine at all, and I install P6 on a Virtual machine, which allows me to run any version at any time. I also have a VM for each customer, so I can back up and/or share all my customers work with my customer by sharing a VM and I am not giving them all my other customers information.

29 Useful Websites

There are many web sites available with lots of great information. I have listed below some that you may wish to visit:

- ❖ **https://www.eh.com.au** my web site for all my books and courses including P6, Microsoft Project and Elecosoft (Asta) Powerproject.

- ❖ **https://www.primavera.com.au** my web site for all my books and courses focusing on P6. This has a page with a list of third-party software that works with P6.

- ❖ **http://www.planningplanet.com** the **Planning Planet** is a site that links networking planning, scheduling, program& project control professionals around the world.

- ❖ **https://www.oracle.com/industries/construction-engineering/** This page is the main portal for information on **Oracle Primavera Construction** software.

- ❖ **https://edelivery.oracle.com** the **Oracle E-Delivery** web site where you may download any Oracle product that is currently available to license.

- ❖ **https://shop.oracle.com** The **Oracle Store** is where you may purchase Oracle software online. You should be able to buy P6 software at a lower price by negotiating with an Oracle Partner.

- ❖ **https://support.oracle.com** The **Oracle Support** page is the main portal for support for Oracle software. You will need to create an account to log into the system and view the support database. This is a link to a site that is an Oracle Primavera site.

- ❖ **https://www.oracle.com/communities/** The Oracle Community site is an Oracle site and includes several links to other Oracle sites such as Blogs, Forums, Wiki, User Groups Support etc.

- ❖ **https://tensix.com/** The **Tensix Consulting** web site has hundreds of great blogs.

- ❖ **https://www.planacademy.com/blog/** The **Plan Academy's** blog has some great articles.

- ❖ **https://ronwinterconsulting.com/** **Ron Winter** has published many articles on P6 and has a wealth of knowledge on the ins and outs of P6.

- ❖ **https://docs.oracle.com/en/** The Oracle Technology web site which has the documentation for Primavera P6 software.

- ❖ **https://projectcontrolsonline.com/** The Project Controls Online.com web site has many P6 blogs.

- ❖ **https://www.planacademy.com/** The Plan Academy, an online training company, produces and delivers online, on-demand training on Project Controls. It also has several good blogs.

If you know of any other website with useful information, please let me know and I will add them to my next publication.

© *Eastwood Harris*